Earthquakes and Acoustic Emission

BALKEMA – Proceedings and Monographs
in Engineering, Water and Earth Sciences

Earthquakes and Acoustic Emission

Edited by
Alberto Carpinteri & Giuseppe Lacidogna
Department of Structural Engineering and Geotechnics, Politecnico di Torino, Italy

CRC Press
Taylor & Francis Group
Boca Raton London New York

CRC Press is an imprint of the
Taylor & Francis Group, an **informa** business
A TAYLOR & FRANCIS BOOK

Cover photographs: left; view of the Astesiano Tower, one of the medieval towers still rising in Alba (an ancient town in Piemonte, Italy), right; a thermographic view of the same tower.

Photos by courtesy of Professor Alberto Carpinteri and Dr. Giuseppe Lacidogna.

CRC Press
Taylor & Francis Group
6000 Broken Sound Parkway NW, Suite 300
Boca Raton, FL 33487-2742

First issued in hardback 2019

Typeset by Charon Tec Ltd (A Macmillan Company), Chennai, India

ISBN-13: 978-0-415-44402-6 (hbk)

Library of Congress Cataloging-in-Publication Data

Earthquakes and acoustic emission / edited by Alberto Carpinteri & Giuseppe Lacidogna.
p. cm.
Includes bibliographical references and index.
ISBN 978-0-415-44402-6 (hardcover : alk. paper)
1. Fracture mechanics. 2. Concrete—Cracking. 3. Buildings—Earthquake effects. 4. Acoustic emission testing. 5. Seismic waves. I. Carpinteri, A. II. Lacidogna, Giuseppe.

TA409.E28 2007
620.1'126—dc22

2007019363

Visit the Taylor & Francis Web site at
http://www.taylorandfrancis.com

and the CRC Press Web site at
http://www.crcpress.com

Contents

Preface

An earthquake is a sudden movement of the ground caused by the release of elastic energy stored in the Earth's Crust. It causes vibrations that propagate outwards from the source as seismic waves. A similar phenomenon occurs – on a different scale – in structural materials under load and is called Acoustic Emission (AE). While the techniques used in seismology have been well developed over the years, the AE technique has not been fully defined. It can be shown, on the other hand, that the AE method can use many seismic analysis techniques, including the localization of the sources or the inversion of source parameters.

The contributions proposed in this book, selected among those presented in Special Sessions at the 11th International Conference on Fracture, Torino (Italy), March 20-25, 2005, aim at providing an outline of the theory and the latest relevant developments on topics related to Earthquakes and Acoustic Emission.

During the last few years, numerous experimental techniques have been employed to evaluate fracture processes, and a number of modelling approaches have been developed to predict fracture behaviour. The non-destructive method based on the Acoustic Emission (AE) technique has proved highly effective, especially to check and measure the damage phenomena that take place inside a structure subjected to mechanical loading.

The acoustic emission is a consequence of micro-cracks forming and propagating in the material and should thus provide an indirect measure of the damage accumulating in the system. For this reason, AE is often used as a non-destructive tool in material testing and evaluation. Beside these practical applications, understanding the statistical properties of AE has become a challenging theoretical problem. The distribution of AE amplitudes follows a power law, suggesting an interpretation in terms of critical phenomena and scaling theories. This behaviour has been observed in several materials such as concrete, masonry and rocks, to name just a few.

On the other hand, another common and familiar example of this behaviour is the occurrence of earthquakes in a particular seismic zone, where the event amplitudes span from the smallest almost imperceptible vibration of the crust to the largest destructive catastrophic events. The AE monitoring technique is similar to the one employed in earthquake control, where seismic waves reach the monitoring stations placed on the surface of the Earth. Though they take place on very different scales, these two families of phenomena – damage in structural materials and earthquakes in geophysics – are very similar: in either case, in fact, we have a release of elastic energy from sources located inside a medium.

The contributions presented in this book have been classified into two categories: Seismic Mechanics and Earthquakes (Part 1), Structural Failure and Acoustic Emission (Part 2), with the aim of bringing together the latest achievements at world-wide level from mechanics of materials to geophysics and presenting the applied (for non-destructive testing and failure evaluation) and theoretical (for critical phenomena in complex systems) implications of earthquakes and acoustic emission monitoring.

Alberto Carpinteri & Giuseppe Lacidogna
Torino, Italy
January 2007

1 Seismic Mechanics and Earthquakes

1.1 Earthquake phenomenology and dynamics

Earthquakes and Acoustic Emission – Carpinteri & Lacidogna (eds)
© 2007 Taylor & Francis Group, London, ISBN 978-0-415-44402-6

THE SCALING OF GEOLOGICAL FAULTS

C. H. Scholz
Lamont-Doherty Earth Observatory
Columbia University, Palisades N.Y., USA

ABSTRACT

Geological faults are shear fractures in rock that may range in length from a cm to 1000 km, allowing for a the study of scaling over an unusually broad range. Their shear displacements are found to scale linearly with fault length, with a proportionality constant of the order of 10^{-2}. Their displacement profiles are self-similar, with linear displacement tapers near the tips. These tip tapers are scale-independent. Faults propagate by forming a brittle process zone in the region surrounding their tips. These consist of intergranular tensile microcracks oriented parallel to the maximum compressive stress in the crack-tip stress field. Their maximum crack density is at the edge of the fault and is scale independent. Crack density falls away exponentially with distance from the fault and the width of the process zone increases linearly with fault length. All of these observations are consistent with an elastic-plastic (CTOA) crack model in which yielding occurs in a volume surrounding the crack tip. This implies that fracture energy increases linearly with fault length such that a classical Griffith type instability will not occur. Similar scaling laws apply to earthquakes and joints (macroscopic opening mode fractures in rock).

1 INTRODUCTION

Geological faults are shear fractures in rock with friction on their interfaces. Work over the past decade has revealed their scaling laws as well as the mechanism by which they propagate. Most of these results, as well as related topics such as fault interactions and populations, has been reviewed in Scholz [1], but because most materials scientists are unlikely to consult that work, we briefly recapitulate the more important results here.

The upper few tens of km of the Earth is brittle and under overall compressive stresses so that deformation is primarily through the formation of faults, which accommodate displacements by frictional sliding in their interiors. The rocks through which they propagate are primarily crystalline aggregates of silicates formed under high temperature and pressure, with sintered grain boundaries. Because three or more silicate phases are usually present, and because each is anisotropic in its material properties, strong stress concentrations are developed at the grain scale that results in the grain boundaries being partially cracked, producing a small but connected crack porosity. This also results in there being a high degree of stress heterogeneity at the grain scale. The engineering materials they most closely resemble are the high toughness ceramics (e.g. Evans [2]).

2 DISPLACEMENT SCALING

Faults initiate at a point and as shear displacement D accumulates in their interiors they grow in length L. Figure 1(left) shows the displacement profiles normalized to fault length for a number of faults in the same rock type with lengths ranging from 690–2200 m. A good data collapse is observed, indicating linear scaling between D and L. Notice also that the displacement taper near the fault tips is approximately linear and is scale invariant. A global collection of data is shown in Figure 1 (right), which shows D scaling linearly with L over seven orders of magnitude. The maximum displacement is shown in that figure: if average displacement were shown, the scaling would be the same with a slightly lower proportionality constant. The proportionality constant is a measure of stress drop, $(\sigma_y - \sigma_f)$ where σ_y is the yield strength of the intact rock and σ_f is the residual friction. The stress drop varies with rock type but is scale invariant and of the order of several hundred MPa. Most faults grow intermittently by earthquakes, a stick-slip frictional phenomena. Earthquake displacement profiles

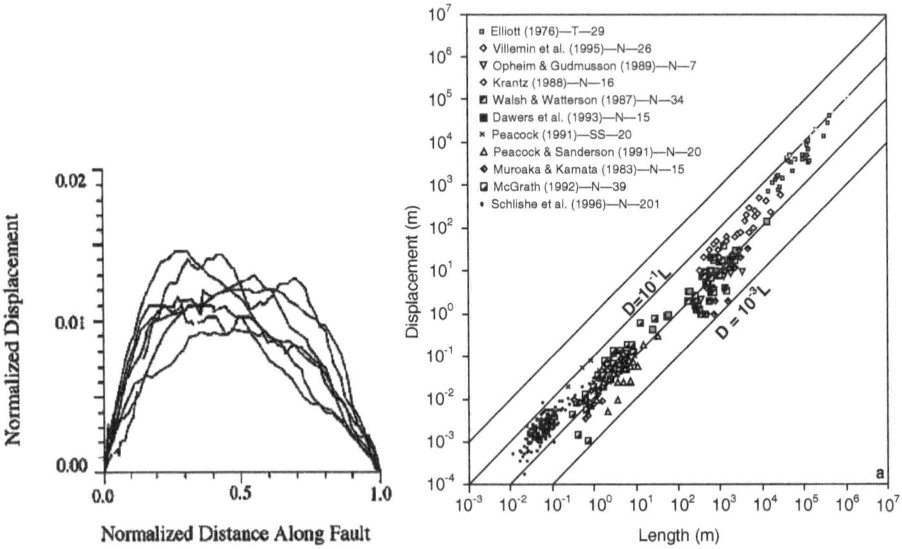

Figure 1: Left, displacement profiles, normalized by fault length, for faults in welded tuff in the Volcanic Tablelands, eastern California (after Dawers [3]). Right, Displacement vs. length for a global fault dataset (after Schlische et al. [4]).

and D/L scaling are identical with faults, but with much smaller stress drops, of the order of 1 MPa. In this case the stress drop represents the drop from static to dynamic friction.

These observations differ markedly with the predictions of linear elastic fracture mechanics. If a fault was an elastic crack, it would have an elliptical displacement distribution, rather than the one observed. Furthermore, if fracture energy G was a constant, we would expect $D \propto \sqrt{L}$. The observed linear scaling of D with L implies that G also scales linearly with L. The crack model that agrees with the fault data is the CTOA (constant tip opening angle) model (Kanninen and Popelar [5]). This is a numerical model in which inelastic yielding is allowed to occur in a volume around the crack tip wherever the stress exceeds the yield strength σ_y. As its name applies, it predicts a scale invariant linear displacement taper near the tip. It also predicts a linear scaling between D and L. Both these agree with the two observations seen in Fig. 1.

3 PROCESS ZONE SCALING

The CTOA model also predicts that inelastic deformation occurs in a volume surrounding the crack tip and that G increases linearly with L. To test these predictions, a series of holes were cored across several strike slip (Mode II) faults in quartzite (a pure sintered quartz rock). Intergranular microcracks were counted from thin sections and their orientations measured (Vermilye and Scholz [6]). A brittle process zone was observed consisting of Mode I microcracks in which the their density fell off exponentially with distance from the faults (Fig. 2, left). The orientations of the poles of the microcracks in the process zone had strong maxima perpendicular to the maximum principal compressive

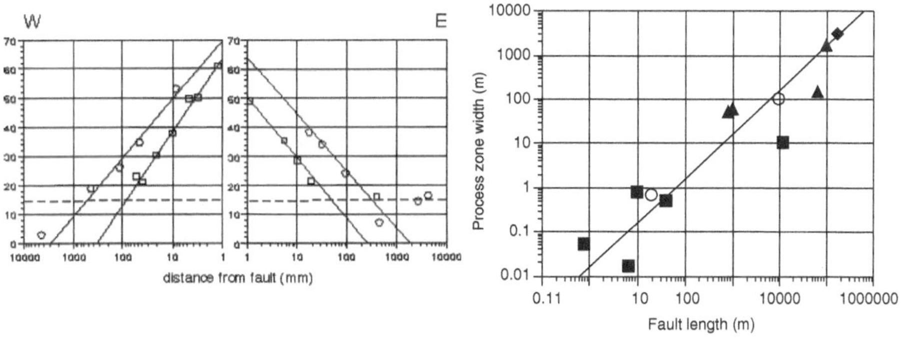

Figure 2: Left, microcrack density as a function of distance from two mode II cracks in quartzite. The dashed line is the background microcrack density. Open circles are for a 40 m long fault, open squares for one 2 m long. Right, process zone widths for faults of different lengths. The slope of the line is 1. Both figures are after Vermilye and Scholz [6].

stress predicted from the Mode II elastic crack tip stress field, whereas those outside the process zone had pole maxima perpendicular to the regional maximum principal compressive stress, which was about 25° from the fault trace, as expected for Coulomb failure with a friction coefficient of about 0.5. The open circles are for a 40 m long fault and the open squares for a 2 m long fault. The dashed line indicates the background microcrack density. The maximum microcrack density is independent of fault length but the width of the process zone is proportional to it. This is shown for a wider range of fault lengths in Figure 2 (right), where the line has a slope of 1. This linear scaling of process zone width with fault length indicates that G scales linearly with L because the surface energy associated with fault growth is the sum of the surface energy of all the microcracks in the process zone.

4 JOINTS AND EARTHQUAKES

Although less work has been done on them, it appears that joints (macroscopic mode I cracks in rock) also best agree with the CTOA model. They have linear scaling between opening and length and tend to have linear opening profiles near the crack tip (Vermilye and Scholz [7]). They have also been observed in the laboratory to have brittle process zones (Swanson [8]). Earthquakes also have linear scaling between D and L and linear tip tapers (Scholz [1]).

5 DISCUSSION

Neither the elastic crack model or LEFM is appropriate for modeling the growth of macroscopic cracks in rock. Because G scales with L rather than being constant, the classic Griffith instability does not occur. The only instability in faults is the frictional one which causes earthquakes. Faults grow intermittently at the time of earthquakes (Cowie and Scholz [9]), but otherwise are quasi-static. We think that the development of brittle process zones at fault and joint tips results from the granular nature of the rock and the resulting stress heterogeneity at the grain scale. It appears that fractures high-toughness ceramics behave the same as rock. They develop brittle process zones and have R-curve behavior (Evans [2]). Although in the laboratory fracture toughness K is measure over only a factor of 2 to 4 and so cannot be proved, from our observations of G scaling with L for rock, we would interpret the R-curve behavior of these ceramics as meaning that K scales with $\sqrt{L}$.

References

[1] Scholz, C. H. The Mechanics of Earthquakes and Faulting, 2nd ed. (Cambridge Univ. Press, Cambridge, 2002).

[2] Evans, A. G. Perspective on the development of high-toughness ceramics. J. Amer. Ceramics Soc. 73, 187–206 (1990).
[3] Dawers, N. H., Anders, M. H. & Scholz, C. H. Growth of Normal Faults – Displacement-Length Scaling. Geology 21, 1107–1110 (1993).
[4] Schlische, R. W., Young, S. S., Ackermann, R. V. & Gupta, A. Geometry and scaling relations of a population of very small rift-related normal faults. Geology 24, 683–686 (1996).
[5] Kanninen, M. F. & Popelar, C. H. Advanced Fracture Mechanics (Oxford Univ. Press, Oxford, 1985).
[6] Vermilye, J. M. & Scholz, C. H. The process zone: A microstructural view of fault growth. J. Geophys. Res.-Solid Earth 103, 12223–12237 (1998).
[7] Vermilye, J. M. & Scholz, C. H. Relation Between Vein Length and Aperture. J. Struct. Geol. 17, 423–434 (1995).
[8] Swanson, P. L. Tensile fracture resistance mechanisms in brittle polycrystals: An ultrasonic and microscopic investigation. J. Geophys. Res. 92, 8015–8036 (1987).
[9] Cowie, P. A., and Scholz, C.H. Growth of faults by accumulation of seismic slip. J. Geophys. Res.97 (B7), 11085–11095 (1992).

Earthquakes and Acoustic Emission – Carpinteri & Lacidogna (eds)
© 2007 Taylor & Francis Group, London, ISBN 978-0-415-44402-6

EFFECTS OF NONLINEAR WEAKENING ON EARTHQUAKE SOURCE SCALINGS

J.-P. Ampuero
Geosciences Department, Princeton University, USA

ABSTRACT

An earthquake is usually modelled as an extension of the Dugdale approach: as in LEFM, fracture energy G_c is completely spent in a relativelly small region close to the crack tip, the process zone; and the fault strength is described by a slip weakening law that reaches a residual level at a characteristic slip D_c. Beyond its regularizing properties, slip weakening is a phenomenological fact of experimental rock mechanics. Recent laboratory observations show that slip weakening can be a persistent process over large amounts of slip, in contradiction to our usual view of a finite D_c. New seismological constraints from seismic nucleation phases and from the scaling of radiated energy with magnitude, seem to point to a similar interpretation. Persistent weakening can also be viewed as a lumped representation of off-fault non linear processes occurring in the wake of the process zone. On this basis, power law weakening laws have been proposed that feature a very steep weakening rate in the short slip range followed by a long tailed, non linear, weakening process with no characteristic slip. On the other hand there is growing interest in the apparent scaling of fault properties, such as D_c and G_c, with earthquake size. This is a key issue in understanding how much can be learned about large destructive earthquakes from the observation of smaller but more frequent ones and from laboratory experiments. I explore here the possible effects of non linear strength drop on macroscopic earthquake source parameters, such as magnitude, rupture size and radiated energy, and their interrelations mainly via numerical simulation of earthquake dynamics under a general family of empirical nonlinear slip weakening laws.

1 INTRODUCTION

In the simulation of earthquake rupture dynamics the linear slip weakening model is the simplest rupture criterion that accounts for a finite fracture energy G_c and regularizes the crack tip stress fields [1]. This allows for a straightforward implementation in numerical codes, which is an argument for his widespread use in earthquake seismology. An additional ingredient, already present in the Dugdale model, is that during an earthquake the strength of the fault drops completely to a residual value beyond a characteristic slip Dc that has been often viewed as a material property. In particular, in the linear slip weakening model the strength drop is linear with slip. In this contribution I will address the effects of relaxing the assumption of complete strength drop by testing an empirical family of non-linear, long-tailed, slip dependent weakening laws. This abstract reviews the experimental, seismological and theoretical motivations for such modelling. The results of the simulations and their analysis will be thoroughly discussed in the oral presentation.

2 EXPERIMENTAL CONSTRAINTS

Slip weakening is generally observed in experimental rock mechanics [2]. In the early 80's a modern description of second order effects in friction was developped, leading to the rate-and-state formalism [3, 4]. Rate-and-state contains the slip-weakening ingredient in some limit and has been successfully applied to slow earthquake initiation, post-seismic relaxation, earthquake clustering and triggering and other long term processes. However, experimental rock mechanics still has to reach some of the important conditions proper for real earthquakes, such as fast slip rates ($\approx$1 m/s) and large displacements ($>$1 m). For this reason, earthquake dynamics simulations are often performed under hypothetical strength weakening laws that enclose a limited set of physical ingredients into a tractable mathematical form. The most fundamental common ingredients are the fracture energy G_c and the characteristic slip D_c. Recently, the experimental frontier of large displacements was explored by shearing samples of granular material in a large rotary apparatus [5]. The granular material simulates the gouge believed to be present in the core of deep active fault zones. As deformation spontaneously localizes at the inner boundary of the sample, the experimental behavior is consistent with our usual view of fault slip. The most original observation was a persistent slip weakening process: the gouge approached residual strength only after slipping several meters. This is in sharp contrast with the usually small values of $D_c \approx 0.1$ mm observed in small scale experiments and even to our usual view of a finite D_c. A more classical rate-and-state behavior with short D_c was also consistently observed but was found to be a relatively minor effect. On this basis, a robust slip weakening law was proposed [5] in the form of an inverse power law:

$$\Delta\, strength \propto (\varepsilon + slip)^{-p} \qquad (1)$$

with exponent p $\approx$ 0.4 and a less well resolved "regularization" slip $\varepsilon < 1$ mm. The distinctive features of this new family of laws are: a very steep weakening rate in the short slip range followed by a long tailed, non linear, weakening process with no characteristic slip. Whereas caution is still required when extrapolating these low velocity experiments to natural earthquake conditions, we will see next that some seismological observations seem to converge to similar interpretations.

3 SEISMOLOGICAL CONSTRAINTS

On the observational seismology side, a major goal is to use recorded seismograms to constrain dynamic fault properties. For large and well instrumented earthquakes, near-field data can now be inverted for the distribution of strength excess, strength drop and D_c [7]. Resolution is problematic due to the limited frequency band of the data. However a more fundamental issue is the strong trade-off between the inverted parameters [8]. This is a signature of the more fundamental role of G_c, which is a compound parameter: the integral of the strength weakening curve up to D_c. Theory indicates that only few aspects of dynamic rupture depend on higher order properties of the weakening law: some seismic nucleation phases (the initial shape of seismograms) depend on the weakening rate [9] and the transition to supershear rupture velocity depends strongly on strength excess [10]. Unfortunately both phenomena are rarely observed, although in the first case optimal conditions should be found in the new generation of deep borehole instrumentation through active faults. So far it seems that we should conform to a minimal description of fault strength in terms of fracture energy G_c.

Moreover most earthquakes are not favorable to the detailed studies of near-field strong-motion seismology, either because of their small magnitude or due to the lack of dense instrumentation. Frequently only macroscopic constraints are available. Among the usual macroscopic quantities two are fundamental and their inter-relations are the subject of ongoing debates: seismic moment M_0 and radiated energy E_r. Whereas the former is routinely reported, estimation of the latter has proved more problematic [11], but much progress has been done recently [12]. Inversion schemes for higher order properties of the seismic moment [13] should soon give further consistent constraints from the far-field on earthquake size, rupture velocity, duration and directivity.

Instead of focusing on very large earthquakes the key is in trying to retrieve clues from as large as possible a magnitude range. An important goal is to bridge the gap between the well observed large earthquakes and the well constrained laboratory scales: how much can be learned about destructive earthquakes from the observation of smaller but more frequent ones and from laboratory experiments? This requires the development of specialized methods to analyse microseismicity, such as the forthcoming data from deep borehole experiments, and an increased synergy between earthquake source seismology and mining and induced hydro-fracture seismology.

In the meantime, the E_r/M_0 scaling debate has been enriched with interpretations based on the slip weakening framework. It was recently shown [14] that a broad magnitude range of $E_r = M_0$ observations was consistent with power law slip weakening. This is analogous to the previously discussed laboratory derived weakening laws, eqn (1).

On the other hand, the hypothesis of linear slip weakening has been tested through new seismological constraints from seismic nucleation phases [9]. In non-LEFM, large scale yielding, models of nucleation under linear weakening, the initial shape of seismograms is predicted to depend on the weakening rate W, through a balance between slip weakening and fault impedance (radiation damping). On a broad magnitude range, the estimated weakening rates were observed to be magnitude dependent, $W \propto M_0^{-1}$. A possible interpretation is non-linear strength weakening $W \propto slip^{-1}$, which points again to the family of laws of eqn (1).

4 THEORETICAL CONSTRAINTS

The steep weakening rate in the short slip range featured by eqn (1) is essential to understand the coexistence of large and small earthquakes. If G_c is a material property LEFM predicts the existence of a minimal stable crack size which, considering the seismological estimates of G_c for large earthquakes, would be in contradiction with the observation of microearthquakes. Scale-dependency of G_c is one way out, but an alternative view comes from a non-LEFM, large scale yielding, analysis of earthquake nucleation. It has been shown [15] that under linear slip weakening and general loading conditions the minimal earthquake nucleation size is related to the weakening rate W. It is reasonably expected that non linear weakening will allow for a broad range of nucleation sizes.

The search for complexity in simulated seismic cycles under deterministic continuum mechanics was elusive until recently [16]. One of the ingredients that were found to be required to generate complex synthetic seismicity (Gutenberg-Richter distribution of earthquake sizes) is the presence of two decoupled slip-scales in the weakening law. It is then reasonable to expect that the slip weakening family eqn (1), with practically no characteristic slip or alternatively with a continuum of slip-scales, will be favorable to the generation of realistic earthquake cycle complexity.

The seismological observations presented in the previous section can alternatively be interpreted as a signature of scale-dependent fault properties. Some have pursued this idea within the framework of fault friction [17], with controversial propositions. It is perhaps more natural to approach the issue from the standpoint of off-fault non linear processes. At the outset of modern dynamic earthquake modelling [1] it was recognized that non linearities of the material surrounding the fault might play an essential role in fault dynamics. A simple yielding bulk rheology was shown to account for usual experimental and seismological facts: low limiting rupture velocities, a moderate fraction of shear wave velocity; and very large apparent fracture energies, the estimates of G_c for large earthquakes being many orders larger than the laboratory estimates. Surprisingly enough, simulation of earthquake dynamics has been since conducted under the more restricted fault friction paradigm. Among the arguments in favor of the friction framework are: the advent of the modern experimentally derived friction laws, some evidence for intense coseismic strain localization and its numerical and analytical tractability. Despite efforts to study the feedback of secondary dynamic faulting on main fault rupture propagation [18], the development of a complete description of non linear off-fault processes is still a rather unexplored frontier of earthquake seismology. In the meantime it should be born in mind that the surface friction model is a lumped representation of fully tri-dimensional dynamic processes. This reconciles the apparent scale-dependency of fault friction parameters: they should not be understood as material properties but as effective properties, just as toughness is understood in fracture mechanics in the presence of *R*-curve behavior [19]. The apparent scale dependency is the result of the dynamic evolution of the tri-dimensional process region, which is expected to grow with rupture size [1].

The development of a dynamic bulk damage or elasto-plastic model is out of the scope of this contribution. I will instead focus on the non linear friction laws of the form of eqn (1). However, non linear weakening can be viewed as a preliminary step to that ultimate goal: it can be seen as a lumped model for non linear processes occuring within the wake of the off-fault process zone.

5 NUMERICAL SIMULATIONS

Spontaneous dynamic rupture is simulated in 2D mode III configuration on a planar fault embedded in a linear elastic unbounded medium. The problem is numerically solved with a spectral boundary integral equation method, including time-truncated kernels to allow for long runs containing slow nucleation stages [20]. The first rupture is nucleated by a slow uniform "tectonic" load on top of a peaked non uniform initial stress that crudely mimmics the state of stress left by previous ruptures [15]. Subsequent events are spontaneously triggered by the "tectonic" load. Instantaneous or time-dependent healing can be included.

I will focus on static and dynamic macroscopic properties of the simulated ruptures such as magnitude/size scalings and magnitude/radiated energy scalings. In the usual seismological interpretation of macroscopic quantities it is implicitly assumed that local quantities, such as stress drop and fracture energy consumption, have weak fluctuations along the fault. This is not a very robust assumption when weakening is strongly non-linear. I will illustrate the possible bias with the help of numerical modelling. This will also shed some light on how to infere weakening properties from far-field observations. The results of the numerical simulations will be reported and discussed during the meeting.

6 CONCLUSION

Experimental and seismological observations are converging towards new paradigms of earthquake dynamics modelling. I suggest that some longstanding open questions of source seismology, specially those related to source scaling relations, may find a way out with the advent of models accounting for off-fault dynamic non linearities and their feedback onto main fault rupture propagation. As a preliminar step I explore the possible effects of an empirical non linear slip weakening law on macroscopic earthquake source parameters. The main motivations have been exposed here. The results of numerical simulations will be reported and discussed during the meeting.

References

[1] Andrews, D. J., "Rupture propagation with finite stress in antiplane strain", J. Geophys. Res., vol. 81, pp. 3575–3582, 1976.

[2] Ohnaka, M. and Shen, L., "Scaling of the shear rupture process from nucleation to dynamic propagation: implications of geometric irregularity of the rupturing surfaces", J. Geophys. Res., vol. 104, pp. 817–844, 1999.

[3] Dieterich, J., "Modeling of rock friction: 1. Experimental results and constitutive equations", J. Geophys. Res., vol. 84, pp. 2161–2168, 1979.

[4] Ruina, A. "Slip instability and state variable friction laws", J. Geophys. Res., vol. 88, pp. 10359–10370, 1983.

[5] Chambon, G., Schmittbuhl, J., and Corfdir, A., "Laboratory gouge friction: seismic-like slip weakening and secondary rate- and state- effects", Geophys. Res. Lett., vol. 29(10), 10.1029/2001GL014467, 2002.

[6] Chambon, G., Schmittbuhl, J., and Corfdir, A., "Frictional response of a thick gouge core: 2. friction laws and implications for faults", J. Geophys. Res., revised, 2004.

[7] Peyrat, S., and Olsen, K. B., "Nonlinear dynamic inversion of the 2000Western Tottori, Japan, earthquake, Geophys. Res. Lett., vol. 31, doi:10.1029/2003GL019058, 2004.

[8] Guatteri, M., and Spudich, P., "What can strong motion data tell us about slip-weakening fault friction laws?", Bull. Seism. Soc. Am., vol. 90, pp. 98–116, 2000.

[9] Ampuero, J.-P., "Etude physique et num'erique de la nucléation des séismes", Ph. D. Thesis, Univ. Paris VII, 2002.

[10] Andrews, D. J., "Rupture velocity of plane strain shear cracks", J. Geophys. Res., vol. 81, pp. 5679–5687, 1976.

[11] Ide, S., Beroza, G. C., Prejean, S. G., and Ellsworth, W. L., "Apparent break in earthquake scaling due to path and site effects on deep borehole recordings", J. Geophys. Res., vol. 108, 10.1029/2001JB001617, 2003.

[12] Venkataraman, A., and Kanamori, H., "Observational constraints on the fracture energy of subduction zone earthquakes", J. Geophys. Res., vol. 109, doi:10.1029/2003JB002549, 2004.

[13] McGuire, J. J., "Estimating finite source properties of small earthquake ruptures", Bull. Seism. Soc. Am., vol. 94(2), 2004.

[14] Abercrombie, R. E., and Rice, J. R., "Can observations of earthquake scaling constrain slip weakening?", submitted to Geophysical Journal International, August 2003.

[15] Uenishi, K., and Rice, J. R., "Universal nucleation length for slip weakening rupture instability under non-uniform fault loading", J. Geophys. Res., vol. 108 (B1), 10.1029/2001JB001681, 2003.

[16] Shaw, B., and Rice, J. R., "Existence of continuum complexity in the elastodynamics of repeated fault ruptures", J. Geophys. Res., vol. 105, pp. 23791–23810, 2000.

[17] Ohnaka, M., "A physical scaling relation between the size of an earthquake and its nucleation zone size", Pure Appl. Geophys., vol. 157, pp. 2259–2282, 2000.

[18] Yamashita, T., "Generation of microcracks by dynamic shear rupture and its effects on rupture growth and elastic wave radiation", Geophys. J. Int., vol. 143, pp. 395–406, 2000.

[19] Lawn, B., "Fracture of brittle solids", 2nd edn, Cambridge Univ. Press, 1993.

[20] Lapusta, N., Rice, J. R., Ben-Zion, Y., and Zheng, G., "Elastodynamic analysis for slow tectonic loading with spontaneous rupture episodes on faults with rate- and state-dependent friction", J. Geophys. Res., vol. 105, pp. 23765–23789, 2000.

Earthquakes and Acoustic Emission – Carpinteri & Lacidogna (eds)
© 2007 Taylor & Francis Group, London, ISBN 978-0-415-44402-6

SEISMIC ENERGY RADIATION FROM DYNAMIC FAULTING

R. Madariaga
Laboratoire de Gèologie, Ecole Normale Supérieure, 4 rue Lhomnod, 75231 Paris Cedex 05, France

ABSTRACT

Seismic radiation allows sismologists to probe the details of the rupture process during an earthquake. Waves recorded in the vicinity of the fault are used to invert for the details of rupture using a mixture of kinematic and quasi-dynamic methods. Seismic radiation is not only a tool to probe the earthquakes,it is also the essential dissipative mechanism by which ruptures exchange energy with the surrounding elastic media. Inversion has been used to study several earthquakes at relatively long wavelength, extending these results to higher frequencies improved understanding of the fundamentals of seismic energy balance which remain obscure, even if the fundamental theorems were established by Kostrov 30 years ago. In this paper we examine the simplest possible earthquake model, a propagating 2D antiplane crack and compute energy balance in detail. We show that energy can be expressed as simple integrals of energy release on the fault, that high frequencies are radaited by fast changes in rupture velocity and in jumps and kinks of the rupture front. We demonstrate that even if energy can be expressed as an integral of slip on the fault, it can not be localized on the fault.

1 INTRODUCTION

In spite of considerable progress in understanding earthquake dynamics, there are still a number of obscure point about elastic energy radiation and its role in controlling rupture. Recent studies have suggested that apparent energy release rate as measure from seismic radiation is of order of several MJ/m^2, a huge number that probably reflects the fact that seismic rupture do not occur on a simple fault surface as modeled in laboratory experiments. Recently, Ide (2002) and Favreau and Archuleta (2003) showed that total energy release rate during the Imperial Valley and Kobe earthquakes was of the same order of magnitude as radiated energy and strain energy released by those earthquakes. This seems to confirm, the claim by Madariaga and Olsen (2001) that earthquakes propagate at an almost critical level so that the Griffith criterion for shear rupture is barely satisfied. This may explain why super-shear rupture is only observed in very special circumstances.

2 RADIATED ENERGY

Kostrov (1964) defined radiated energy as:

$$E_S = -\int_0^\infty \mathrm{d}t \int \dot{u}_i \Delta\sigma_{ij} n_j \,\mathrm{d}S \tag{1}$$

integrated over a surface S that surrounds the fault VERY far from it. Here $\Delta\sigma$ is the tensor of stress change produced by seismic source, $\dot{u}$ is the particle velocity in the far field of the source. So the effectively one can use the far field in it.

The correct expression for radiated energy in terms of fault slip and fault stress was given by Eshelby (1969) and Kostrov (1974), and has been by Madariaga (1976), Ide (1992), Freund (1989), Favreau et al (2003) and many others. Kostrov's expression is:

$$E_S = \frac{1}{2}\int \Delta\sigma_{ij}\Delta u_i n_j \,\mathrm{d}S - \int_\Sigma G\,\mathrm{d}S + \int_0^\infty \mathrm{d}t \int_\Sigma \dot{\sigma}_{ij}\Delta u_i n_j \,\mathrm{d}S \tag{2}$$

where E_s is the total radiated energy, Σ is the fault surface and $\vec{n}$ is the fault normal, Δu_i is the ith component of the slip vector across the fault and $\Delta\sigma$ is the stress tensor change produced by slip. This expression is remarkable in that it expresses total radiated energy as a function only of changes in the state of stress and inelastic slip.

Very often this equation in the seismological literature this equation is integrated by parts to obtain:

$$E_S = \frac{1}{2}\int \Delta\sigma_{ij}\Delta u_i n_j \,\mathrm{d}S - \int_\Sigma \gamma\,\mathrm{d}S + \int_0^\infty \mathrm{d}t \int_\Sigma \sigma_{ij}\Delta\dot{u}_i n_j \,\mathrm{d}S \tag{3}$$

where σ is the TOTAL stress acting on the fault. Although they are equivalent, equations (2) and (3) are very different and should not be used simultaneously in evaluating seismic radiation. In expression (3) the first term is the TOTAL strain energy change while in (2) the first term is just the self-energy change. Eshelby was the first to show that since the radiated field does not depend on absolute values of stress, radiated energy should not depend on absolute stresses either.

Recently Ide (2002) and Favreau and Archuleta (2003) used a version of (2) in order to compute the radiation from several earthquakes for which slip on the fault had been computed by inversion of near field seismograms. In these inversions. Although this inversion technique is only valid at relatively low frequencies, the slip distributions are robust. From the slip distribution a stress change field is computed so that expression (2) can be computed term by term. Many authors have expressed doubts that one can decompose E_s into individual contributions by fault elements. We will examine this problem in detail in the following.

3 RADIATION FROM A SIMPLE ANTIPLANE FAULT

For simplicity, I will study the semi-infinite mode III crack shown on Fig. 1 that extends along the negative x-axis and is in equilibrium for $t < 0$. At $t = 0$ rupture starts moving at constant speed v_r along the positive x axis. The stress intensity factor at the rupture tip at $x = 0$ is noted K_0 and the corresponding quasi-static energy release rate is $G_0 = K_0^2/(2\mu)$.

Radiation from the semi-infinite fault can be computed by Kostrov's (1966) solution for a flat crack (see, e.g. Madiaraga, 1977, 1984) or Adda-Bedia and Arias (2004) who studied the more complex problem of radiation from a fault that has a kink at the origin.

From Madariaga (1977) we know that rupture of semi-infinite fault produces a shear wave front with a particle velocity jump:

$$\dot{u}\left(r,\theta,t\right) = \frac{K_0}{\mu}\frac{v_r}{\sqrt{2\pi r}}\frac{\sin\theta/2}{1 - v_r/\beta\cos\theta}H\left(t - r/\beta\right) \tag{4}$$

at an observer situated at coordinates r, θ. Here μ is the elastic constant an β the shear wave speed. This is what seismologists call a far field solution, i.e. the

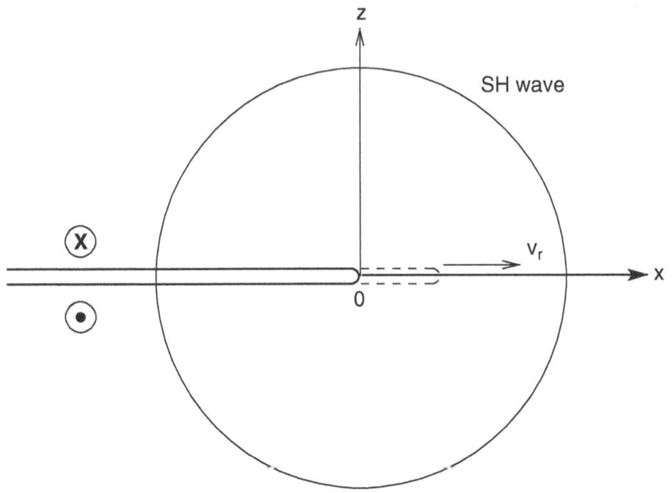

Figure 1: Simple model of a planar antiplane fault that suddenly extends along the x-axis at constant speed v_r. This is one the rare crack radiation problems that can be fully solved exactly.

dominating term as $r \rightarrow \infty$. Since particle velocity is constant behind the rupture front in (2), the total energy radiated by this fault is unbounded and the computation of seismic energy ad defined by Kostrov does not make sense.

4 RADIATION FROM A FINITE MODE III FAULT

In order to compute seismic energy we let the fault of section (1) run for a short length L along the x axis and then stops. When it stops it radiates a wave of the same form a (2) but with r replaced by Fraunhoffer's approximation:

$$r' = \sqrt{\left(x + L\right)^2 + y^2} \approx r\left(1 + L/r\right)$$

The far field wave generated by the finite rupture episode is then:

$$\dot{u}\left(r,\theta,t\right) = \frac{K_0}{\mu}\frac{v_r}{\sqrt{2\pi r}}\frac{\sin\theta/2}{1 - v_r/\beta\cos\theta}$$
$$\left\{H\left(t - r/\beta\right) - H\left[\left(t - r/\beta - L/v\left(1 - v_r/\beta\cos\theta\right)\right)\right]\right\} \quad (5)$$

Using (1) we can compute radiated energy exactly for this model. We get:

$$E_S = \int_0^\infty \mathrm{d}t \int_S \frac{\mu}{\beta}\dot{u}^2\,\mathrm{d}S \quad (6)$$

which gives after a change in the order of integration and use of (5):

$$\begin{aligned} E_S &= \frac{K_0^2}{\mu}\frac{v_r}{\beta}\frac{1}{2\pi}\int_0^{2\pi}\frac{\sin^2\theta/2}{1 - v_r/\beta\cos\theta} \\ &= \frac{K_0^2}{2\mu}L\left[1 - \sqrt{\frac{1 - v_r/\beta}{1 + v_r/\beta}}\right] \\ &= G_0 L - G\left(v_r\right)L \end{aligned} \quad (7)$$

Radiated energy is the difference between the quasi-static and dynamic work of the stress concentration K_0 as it moves a distance L along the x axis.

We can now generalize (7) and write that the total energy for an arbitrarily moving rupture is

$$E_S = \int_0^L \left(G_0\left(x\right) - G\left(v,x\right)\right)\mathrm{d}x \quad (8)$$

a similar result was obtained by Freund (1989) by a modification of Kostrov's equation (2).

5 FAULT SLIP AND ENERGY DISTRIBUTION

The most important question from a seismological point of view is whether (8) or one of its variations can be used to back trace the source of energy from a seismogram to the fault. We would like to know whether the energy observed at time t at the observer comes from an specific point on the fault. For displacement and particle velocity, this problem was already studied in some detail by Bernard and Madariaga (1984) in the 3D case and by Madariaga (1984) in 2D. Following these authors the field observed at the time t at an observer situated at the cylindrical coordinates (r,θ) comes from an isochrone, a curve on the fault for 2D faults or a few isolated point for 1D fault. If the latter problem can be solved, then the computation of energy flow as a function of time and position can be computed.

Unfortunately, radiated energy can not be decomposed in the same way, because radiated energy is averaged on the focal sphere and arrival times depend on the azimuthal angle and distance. Thus, the computations by Ide (2002) and Favreau and Archuleta (2003) can be considered as reasonable estimates of the contribution of points on the fault to total energy balance, but they can not be decomposed into individual contributions of different segments of the fault, unless rupture history is well resolved. The main reason is that fault radiation is not simply proportional to slip, it includes effects of diffraction (and refraction for in-plane problems) by the tip of the fault and other geometrical or mechanical discontinuities.

6 GENERALIZATION

We presented above the solution for the motion of straight antiplane shear fault. Similar results can be obtained for plane shear using the solution by Madariaga (1977) although finding an expression similar to (7) is much more difficult. Recently Adda-Bedia and Arias (2004) have found a general solution to the problem of Fig. 1 where the fault starts moving along kinked fault. Using their results Adda-Bedia and Madariaga (in preparation) found that the expression (7) is still valid if the stress intensity factor is replaced by a modified value that takes into account the effect of the kink on the stress field in the vicinity of the crack tip. Extension of these results to 3D ruptures is much more difficult, except when rather strong kinematic assumptions are made as in the case of a circular crack propagations at constant speed (Madariaga, 1976; Ide, 2002).

7 CONCLUSIONS

We have shown that seismic radiation can be computed exactly from the slip change that occurs on the fault. There are several important caveats that have to be carefully considered, but in principle at least the computation is feasible. The main difficulty is not so much computing energy, but identifying the place on the fault where energy comes at certain time. This requires solving a non-linear back-projection problem when rupture velocity is known for every point on the fault: In reality, determining the rupture velocity is much more difficult problem that computing the distribution of slip, so that I expect that much better rupture invert-ions will be required before we can fully identify the origin of radiation on the fault.

References

[1] Adda-Beida, M., J. Mech. Phys. Solids, 52, 1407–1420, 2004.
[2] Adda-Bedia and R. Arias, J. Mech. Phys. Solids, 51, 1287–1304, 2003.
[3] Eshelby, J.D., J. Mech. Phys. Solids, 17, 177–197, 1969.
[4] Favreau, P. and Archuleta, R., Geophys. Res. Lett., 30, 1198, 2003.
[5] Freund, L.B., Dynamic Fracture Mechanics, Cambridge Univ. Press, New York, 1989.
[6] Ide, S., Bull. Seismol. Soc. Am., 92, 2994–3005, 2002.
[7] Kostrov, B.V., Appl. Math. Mech., 30, 1241–1248, 1966.
[8] Kostrov, B.V., Izv. Earth Physics, 1, 23–40, 1974.
[9] Madariaga, R., Bull. Seismol. Soc. Am., 66, 639–666, 1976.
[10] Madariaga, R., Geophys. J. Roy. Astr. Soc., 51, 625–651, 1977.
[11] Madariaga, R., Annales Geophysicae, 1, 17–23, 1983.
[12] Madariaga, R. and Olsen, K.B. Pageoph., 157, 1981–2001, 2000.

Earthquakes and Acoustic Emission – Carpinteri & Lacidogna (eds)
© 2007 Taylor & Francis Group, London, ISBN 978-0-415-44402-6

DAMAGE RHEOLOGY MODEL AND DECAY LAW OF AFTERSHOCK ACTIVITY

V. Lyakhovsky[1] & Y. Ben-Zion[2]
[1]The Geological Survey of Israel, Jerusalem, 95501, Israel
[2]Department of Earth Sciences, University of Southern California Los Angeles, CA, 90089-0740, USA

ABSTRACT

Aftershocks are the response of a damaged rock surrounding large earthquake ruptures to the stress perturbations produced by the large events. Lyakhovsky et al. (1997a) developed a damage rheology model that provides a quantitative treatment for macroscopic effects of evolving distributed fracturing with local crack density represented by a damage state variable. A thermodynamically-based equation for damage evolution accounts for both degradation and healing of rock as a function of on-going deformation. The governing material properties are constrained by analyses of stress-strain and acoustic emission laboratory data during deformation leading to brittle failure of rocks. For analysis of aftershocks, we consider the relaxation process of a material following the application of a strain step associated with the occurrence of a main shock. The coupled differential equations governing the damage evolution and stress relaxation processes are written in non-dimensional form by scaling the elastic stress to its initial value and the time to characteristic time of damage evolution τ_d. With this, the system behavior is controlled by the single non-dimensional ratio R representing the ratio between damage time scale to the Maxwell relaxation time ($R = \tau_d/\tau_M$). For very small R there is no relaxation and the response consists of constant elastic strain leading to constant rate of damage increase until failure. For very large R there is rapid relaxation without significant change to the level of damage. For intermediate cases the equations are strongly coupled and nonlinear. The analytical solution for the damage evolution can be fitted well for various values of R with a power law similar to the modified Omori law for aftershocks. This also holds for 3-D numerical simulations of aftershock sequences in a multi-layered lithosphere model. Analytical and numerical results suggest that high values of R, corresponding to low viscosity material, produce diffuse aftershock sequences, while low values of R, corresponding to more brittle material, produce clear aftershock sequences.

1 INTRODUCTION

Rocks exhibit a wide variety of rheological behaviors ranging from ductile plastic flow and viscoelastic deformation in the earth mantle and lower crust, to fracture processes controlling the mechanical response and stability of rock mass in the seismogenic zone. A great challenge of theoretical geodynamic studies is to incorporate the interaction between mantle and lower crust into models that simulate deformational processes in the upper crust. Fundamental nonlinear aspects of rock deformation, such as microcrack and flaw nucleation and development of process zones at rupture tips do not have at present accepted quantitative theoretical framework. These aspects are of crucial importance for evolutionary self-organization of faults at various spatio-temporal domains. Lyakhovsky et al. (1997a) developed a thermodynamically-based version of damage rheology, which holds a potential for providing a framework for understanding realistic processes of rock deformation such as those mentioned above. The theoretical results are written in terms of macroscopic variables that can be measured directly in the laboratory. Ben-Zion et al. (1999), Lyakhovsky et al. (2001) and Ben-Zion & Lyakhovsky (2002) showed that the above damage rheology model can be used to understand deformation associated with large earthquake cycles, evolving fault geometries, frequency-size statistics of earthquakes, accelerated seismic release and more.

In the flowing sections we briefly list some manifestations of distributed damage in natural rocks and rock samples which form the experimental basis for our theoretical damage approach. Then we present the main equations of the damage rheology model. Finally we give a 1-D analytical solution and 3-D numerical simulations that indicate a decay of seismic activity after brittle failures in a form similar to the modified Omori law for aftershocks.

2 DISTRIBUTED DAMAGE IN ROCKS

Pioneering studies of fractures and faults treated the crust as an infinite, perfectly elastic medium (e.g., Anderson, 1951). Subsequent studies accounted for the finite length of faults, and the perturbation to the regional stress field due to the proximity of additional faults (e.g., Chinnery, 1966). Field mapping often shows that the density of faults depends on the scale of the map, so higher resolution increases the number of faults in a given domain (Scholz, 1990). This complexity limits the use of methods that specify the positions of isolated cracks in fault zones. Field observations suggest that the size of the process (damage) zone grow with the size of the fracture, in violation of the premises of the critical stress intensity factor approach (Scholz et al., 1993; Vermilye and Scholz, 1998). This is decisively documented around dikes that form by the injection of magma into fractures (Delaney et al., 1986; Weinberger et al., 1995, 2000; Hoek, 1995; Rubin, 1995).

Damage in the form of microcracks profoundly affects rock strength and rock elastic parameters (Nishihara, 1957; Zoback and Byerlee, 1975; Schock, 1977; Schock and Louis, 1982; Lockner and Byerlee, 1980; Alm et al., 1985; Reches and Lockner, 1994; Pestman and Munster, 1996) and leads to vanishing elastic moduli at large stresses just before failure (e.g., Lockner and Byerlee, 1980; Lockner et al., 1992; Hamiel et al., 2004). The investigations of fracturing of rocks (Yukutake, 1989; Lockner et al., 1991; Reches and Lockner, 1994) show that this process can not be described in terms of singlecrack propagation. While linear elastic fracture mechanics assumes the size of the inelastic zone at the crack tip to be negligibly small the experiments show that this zone has a significant size. The finite size effect of the fracture process zone is often treated with models which specify a cohesive zone near the crack tip within the plane of the crack (Dugdale, 1960; Barenblatt, 1962, 1996; Ida, 1972; Palmer and Rice, 1973; Rubin, 1993; Willemse and Pollard, 1998). This approach eliminates an unphysical crack-tip singularity and is useful when the crack geometry is well defined. In most engineering and rock-like materials a propagating crack is preceded by an evolving damage zone around its tip (e.g., Bazant and Cedolin, 1991; Lockner et al., 1991), which controls the macro-crack trajectory and growth rate (Huang et al., 1991; Chai, 1993; Zietlow and Labuz, 1998). Thus it is desirable to account explicitly for the distribution of damage in simulations of fault evolution.

3 THEORY

We briefly present here the physical basis of continuum damage mechanics. Detailed explanations and comparisons with rock mechanic experiments may be found in Lyakhovsky et al. (1997a) and Hamiel et al. (2004). The damage parameter provides a measure of the effect that distributed microcracks have on the macroscopic response of rocks on applied deformation. A volume-averaged measure of damage represents the density of microcracks at a length scale much larger than the length of a typical crack, yet much smaller than the linear size of the volume considered. In a specimen deformed in a rock-mechanics experiment, microcrack density can be measured over a sub-centimeter scale where it is considered as distributed damage. On the other hand, at the crustal length scale, individual faults shorter than the dominant through-going structure can be considered as distributed damage. For any system with a sufficiently large number of cracks, one can define a representative volume in which the crack density is uniform. Following Kachanov (1986) the damage variable is related to the reduction of the elastic moduli of a spatial domain relative to those of an ideal crack free solid. Rabotnov (1988) related the damage variable to a reduction of the effective cross-section area that supports the load. Fiber-bundle models of damage (Newman and Phoenix, 2001; Turcotte et al., 2003) share the same idea, where cracks are equivalent to torn fibers.

The present damage rheology model (Lyakhovsky et al., 1997a) treats two aspects of the physics of damage: (1) A mechanical aspect – sensitivity of the macroscopic elastic shear modulus to distributed cracks and to the sense of loading, and (2) a kinetic aspect – the evolution of damage (degradation/recovery of elasticity) in response to loading. The cumulative effect of distributed micro-cracks and flaws in the elastic material leads to non-linearity, which is described by an energy potential equation of the form:

$$U = \frac{1}{\rho}\left(\frac{\lambda}{2} I_1^2 + \mu I_2 - \gamma I_1 \sqrt{I_2}\right), \tag{1}$$

where $I_1 = \varepsilon_{kk}$ and $I_2 = \varepsilon_{ij}\varepsilon_{ij}$ are two strain invariants of the strain tensor ε_{ij}. This is the simplest mathematical expression for the elastic energy that expands the classical form with two Lame constants λ and μ, being a non-analytical second order function of two strain invariants I_1 and I_2 and additional modulus γ. Lyakhovsky et al. (1997b) show that non-linear elastic deformation of rocks including dilation is fitted well with stress-strain relations derived from the potential (1). In opposite to the Murnaghan (1951) expression, who added a higher order term to the elastic energy, the present approach preserves its non-linear features even for small deformation.

The kinetic aspect of the damage rheology model is introduced by making the Lame parameters λ, μ and the additional modulus γ functions of the damage parameter $0 \leqslant \alpha \leqslant 1$. The amount of damage α evolves in time as a result of an applied load. Using the balance equations of the energy and entropy, and accounting for irreversible changes related to viscous deformation and material damage, the equation of damage evolution has the form (Lyakhovsky et al., 1997a):

$$\frac{d\alpha}{dt} = -C\frac{\partial U}{\partial \alpha}, \tag{2}$$

where the positive constant or function of state variables C provides the non-negative local entropy production related to damage evolution. This approach describes not only damage increase, but also a process of material recovery associated with healing of microcracks, which is favored by high confining pressure, low shear stress and high temperature. Agnon and Lyakhovsky (1995) chose the moduli μ and γ to be linear functions of α and modulus λ be constant. Later analysis of laboratory acoustic emission and stress-strain data (Hamiel et al., 2004) show that the quality of data fitting may be improved assuming power law relation between damage and elastic moduli. However, a linear relation is still very reasonable approximation for conditions in the seismogenic zone. Increasing the third modulus γ from zero for damage free Hookean material to its maximum value at critical damage amplifies the non-linearity of rock elasticity with damage accumulation. Using the elastic energy (1) the equation for kinetic of damage accumulation is (Lyakhovsky et al., 1997a):

$$\frac{d\alpha}{dt} = C_d I_2(\xi - \xi_0) \quad \text{for} \quad \xi \geq \xi_0, \tag{3}$$

where $\xi = I_1/\sqrt{I_2}$ is the strain invariant ratio. Equation (3) includes a parameter ξ_0 that indicates the transition stage from strengthening to degradation. Agnon and Lyakhovsky (1995) and Lyakhovsky et al. (1997a) related this parameter to the angle of internal friction by considering the critical shear stress for Mohr-Coulomb sliding. They received $\xi_0 = -0.8$ for typical ratios of elastic moduli for damage free material $\lambda/\mu \sim 1$ (Poisson's ratio of 0.25) and internal friction angle $\sim 40°$ (Eq. 37 and Fig. 3 of Lyakhovsky et al., 1997a). This value varies little for different rocks with Poisson ratios between 0.2 and 0.3 and is used for the following up numerical simulations. The parameter C_d is the damage rate constant for positive damage evolution (degradation), which defines the time needed to achieve failure after the onset of damage at $\xi = \xi_0$. C_d is assumed to be a material property and its value has been estimated to vary from 0.5 to 5 s^{-1} for different rocks tested at more than 20 MPa confining pressure and room temperature (Lyakhovsky et al., 1997a). Detailed recent analysis of laboratory data (Hamiel et al., 2004) also shows that with the onset of acoustic emission and damage evolution, there is non-reversible deformation before the final macroscopic failure. This can be accounted for using a damage-related viscosity η, which for positive rate of damage is:

$$\eta = \frac{\eta_0}{\dot{\alpha}}. \tag{4}$$

The viscosity value increases with decreasing damage rate and is infinity large (no any irreversible deformation) for negative rate of damage (healing). The damage-related viscosity accounts for irreversible deformations at the length scale below the length scale of a representative volume for which the average damage parameter was defined.

4 DECAY OF ACTIVITY (ANALYTICAL SOLUTION)

The fully 3-D model formulation is significantly simplified in a one-dimensional case. In spite of the non-linearity of the general stress-strain relations based on the elastic potential (1), a corresponding 1-D stress-strain still has a linear form:

$$\sigma = 2\mu_0(1 - \alpha)\varepsilon, \tag{5}$$

with the damage parameter α evolving according to the 1-D version of the damage rate equation (2):

$$\dot{\alpha} = C_d \varepsilon^2. \tag{6}$$

Equations (5, 6) describe evolving elastic strain and damage as a result of ongoing deformation. Equation (6) differs from the 3-D equation (3) by a factor $(\xi - \xi_0)$, which relates the rate of damage increase with the 3-D structure of the deformation. Viscous relaxation gradually reduces the deviatoric component of the elastic strain leading to decreasing the ξ value. When the factor $(\xi - \xi_0)$ approaches zero, the rate of damage evolution decreases and the system achieves a static state. The fully coupled evolution is simulated numerically and discussed in the next section. Here we assume that the factor $(\xi - \xi_0)$ is of the order of one, which allows analytical analysis of the system behavior. The irreversible strain component before the final macroscopic failure is given by:

$$\sigma = 2\eta e, \tag{7}$$

where e is the rate of gradual irreversible strain and the viscosity η is defined in equation (4). The set of equations (4–7) governs the 1-D behavior of the system. A similar set of equations with constant viscosity, excluding eq. 4, was previously analyzed by Lyakhovsky et al. (1997a) to describe a stickslip motion under a constant rate of applied deformation. The solution of equations (5, 6) without irreversible strain accumulation prior to the final brittle failure gives a power-law time-to-failure relation consistent with the power-law seismic activation prior to some large earthquakes (Ben-Zion and Lyakhovsky, 2002; Turcotte et al., 2003). The analytical power-law scaling was obtained for a constant applied stress that mimics natural boundary conditions for a system at the stage of accelerating seismic release. This result was confirmed by numerical simulations in a model of rheologically layered lithosphere governed by damage rheology (Ben-Zion and Lyakhovsky, 2002). Here we analyze the system behavior with constant total strain boundary condition. Such a condition is appropriate for a region where aftershock series are collected after a strong event.

We scale the elastic strain to its initial value ε_0 and time to the damage time scale $\tau_d = 1/C_d\varepsilon_0^2$. The analysis indicates that the system behavior is defined by only one parameter R, the ratio between τ_d and Maxwell time-scale for viscous relaxation τ_M. The latter is:

$$\tau_M = \frac{\eta}{\mu} \cong \frac{\eta_0}{\mu_0 \dot{\alpha}} \cong \frac{\eta_0}{\mu_0 C_d \varepsilon_0^2}, \tag{8}$$

and the ratio R is:

$$R = \frac{\tau_d}{\tau_M} \cong \frac{\mu_0}{\eta_0}. \tag{9}$$

After elimination the elastic strain and integration, the damage evolution is described by the equation:

$$\operatorname{erf}\left(\sqrt{\frac{R}{2}}\right) - \operatorname{erf}\left(\sqrt{\frac{R}{2}}(1-\alpha)\right) = \varepsilon_0^2 \exp\left(-\frac{R}{2}\right)\sqrt{\frac{2R}{\pi}} \cdot t \tag{10}$$

The solution (10) shows that the system does not follow a power-law scaling. However for a given R-value, the calculated rate of damage can roughly be approximated by a power-law function similar to that used for the Omori law. The solution (10) can also be used to estimate the time needed for α to grow from 0 to 1, which roughly increases as exp (R). Thus, we expect a faster decay of the fracture process corresponding to shorter aftershock series for small R-values, and long-lasting series or even swarms of events for high R-values.

4 DECAY OF ACTIVITY (3-D NUMERICAL SIMULATIONS)

We present results of 3-D numerical simulations of the coupled evolution of earthquakes and faults in a model consisting of a seismogenic upper crust governed by damage rheology over a viscoelastic layered substrate. Ductile flow in the lower crust and upper mantle is governed by power-law temperature-dependent rheology. Our three dimensional numerical simulations use the Fast Lagrangian Analysis of Continua (FLAC) algorithm (Cundall and Board, 1988; Cundall, 1989; Poliakov et al., 1993). The formulation is explicit-in-time, using an updated Lagrangian scheme to provide the capability for large strains. The FLAC algorithm offers advantages over conventional finite element schemes in case where material instability occurs. Physical instability is modeled without numerical instability if inertial terms are included in the equilibrium equations.

The simulated aftershock statistics depend on the parameter R responsible for the stress relaxation in the seismogenic zone and chosen temperature gradient. The results indicate that model realizations with low R-values and low temperature gradient (relatively cold lower crust) has clear aftershock record. In contrast, stress relaxation is very fast in model realizations with high R-values and hot lower crust, producing diffuse swarm-like patterns. The analyses of the synthetic aftershock catalogs show that the temporal aftershock distribution can be fitted well for various values of R with a power law similar to the modified Omori law.

References

[1] Agnon, A. & Lyakhovsky ,V., 1995. Damage distribution and localization during dyke intrusion. In: The physics and Chemistry of Dykes. Eds. Baer, G. & Heimann A., pp. 65–78, Balkema. Brookfield.

[2] Alm, O., Jaktlund, L.L. & Kou, S., 1985. The influence of microcrack density on the elastic and fracture mechanical properties of Stripa granite. Phys. Earth. Planet. Inter., 40, 161–179.

[3] Anderson, E.M. The dynamics of faulting, 183 pp., Oliver and Boyd, London, 1951.

[4] Barenblatt, G.I., 1962. Mathematical theory of equilibrium cracks in brittle fracture. Acv. Appl. Mech., 7, 55–129.

[5] Barenblatt, G.I., 1996. Scaling, self-similarity, and intermediate asymptotics, Cambridge Univ. Press.

[6] Bazant, Z.P. & Cedolin, L., 1991. Stability of structures. Elastic, Inelastic, fracture and damage theories, Oxford University Press.

[7] Ben-Zion, Y. & Lyakhovsky, V., 2002. Accelerating seismic release and related aspects of seismicity patterns on earthquake faults. Pure and Appl. Geophys. 159, 2385–2412.

[8] Ben-Zion, Y., Dahmen, K., Lyakhovsky, V., Ertas, D. & Agnon, A., 1999. Self-driven mode switching of earthquake activity on a fault system. Earth Planet. Sci. Lett. 172, 11–21.

[9] Chai, H., 1993. Observation of deformation and damage at the tip of cracks in adhesive bonds loaded in shear and assessment of a criterion for fracture. Int. J. Fracture, 60, 311–326.

[10] Chinnery, M.A., 1966. Secondary faulting. Canad. J. Earth Sci., 3, 163–190.

[11] Cundall, P. A., 1989. Numerical experiments on localization in frictional materials. Ign. Arch., 59:148–159.

[12] Cundall, P. A. & Board, M., 1988. A microcomputer program for modeling large-strain plasticity problems. Numerical Methods in Geomechanics, Proceedings of 6th International Conference on Numerical Methods in Geomechanics, Innsbruck, edited by C. Swoboda, pp. 2101–2108, R.A. Balkhema, Brookfield, Vt.

[13] Delaney, P.T., Polard, D.D., Ziony, J.I., & McKee, E.H., 1986. Field relations between dikes and joints: emplacement processes and paleostress analysis. J. Geophys. Res., 91, 4920–4938.

[14] Dugdale, D.S., 1960. Yielding of steel sheets containing slits. J. Mech. Phys. Solids, 8, 100–104.

[15] Hamiel, Y., Liu, Y., Lyakhovsky, V., Ben-Zion, Y. & Lockner, D., 2004 (in press). A visco-elastic damage model with application to stable and unstable fracturing. Geophys. J. Int.

[16] Hoek, J.D., 1995. Dyke propagation and arrest in Proterozoic tholeitic dyke swarms, Vestfold Hills, East Antarctica, in The Physics and Chemistry of Dykes, eds. Baer & Heimann, Balkema, Rotterdam, 79–95.

[17] Huang, W-L., Kunin, B. & Chudnovsky, A., 1991. Kinematics of damage zone accompanying curved crack. Int. J. Fracture. 50, 143–152.

[18] Ida, Y., 1972. Cohesive force across the tip of longitudinal shear crack and Griffith's specific surface energy, J. Geophys. Res., 77, 3796–3805.

[19] Kachanov, L.M., 1986. Introduction to continuum damage mechanics, pp. 135, Martinus Nijhoff Publishers.

[20] Krajcinovic, D., 1996. Damage Mechanics. Elsevier, Amsterdam.

[21] Lockner, D.A. & Byerlee, J.D., 1980. Development of fracture planes during creep in granite. 2nd Conference on Acoustic Emission/Microseismic Activity in Geological Structures and Materials, Trans-Tech. Publications, Clausthal-Zellerfeld, Germany.

[22] Lockner, D.A., Byerlee, J.D., Kuksenko, V., Ponomarev, A. & Sidorin, A., 1991. Quasi-static fault growth and shear fracture energy in granite. Nature, 350, 39–42.

[23] Lockner, D.A., Byerlee, J.D., Kuksenko, V., Ponomarev, A. & Sidorin, A., 1992. Observations of quasi-static fault growth from acoustic emission. Fault mechanics and transport properties of rocks, International geophysics series, v. 51, B. Evans and T.-F. Wong (eds.), Academic Press, 3–31.

[24] Lyakhovsky, V., Ben-Zion, Y. & Agnon, A., 1997a. Distributed damage, faulting and friction. J. Geophys. Res. 102, 27, 635–27, 649.

[25] Lyakhovsky, V., Ben-Zion, Y. & Agnon, A., 2001. Earthquake cycle, fault zones and seismicity patterns in a rheologically layered lithosphere. J. Geophys. Res. 106, 4,103–4,120.

[26] Lyakhovsky, V., Reches, Z., Weinberger, R. & Scott, T.E., 1997b. Nonlinear elastic behavior of damaged rocks, Geophys. J. Int., 130, 157–166.

[27] Lyakhovsky, V.A. & Myasnikov, V.P., 1985. On the behavior of viscoelastic cracked solid. Physics of Solid Earth, 4, 28–35.

[28] Murnaghan, F.D., 1951. Finite deformation of an elastic solid, John Wiley, Chapman, New York.

[29] Newman, W.I. & Phoenix, S.L., 2001. Time-dependent fiber bundles with local load sharing. Physical Rev. E. 63 (2).

[30] Nishihara, M., 1957. Stress-strain relation of rocks, Doshisha Eng. Rev., 8, 32–54.

[31] Palmer, A.C. & Rice, J.R., 1973. The growth of slip surfaces in the progressive failure of over-consolidated clay. Proc. Roy. Soc. London, Ser. A, 332, 527–548.

[32] Pestman, B.J. & Munster, J.G., 1996. An acoustic emission study of damage development and stress – memory effects in sandstone". Int. J. Rock Mech. Min. Sci., 33, 585–593.

[33] Poliakov, A., Cundall, P., Podladchikov, Yu. & Lyakhovsky, V., 1993. An explicit inertial method for the simulation of viscoelastic flow: An evaluation of elastic effects on diapiric flow in two and three layer models. Proceeding of NATO Advanced Study Institute on Dynamic Modeling and Flow in the Earth and Planets. In: D.B. Stone and S.K. Runcorn (eds.),

Flow and Creep in the Solar system: Observations, Modeling and Theory, Kluwerr, Holland, 175–195.

[34] Rabotnov, Y.N., 1988. Mechanics of deformable solids, pp. 712, Moscow, Science.

[35] Reches, Z. & Lockner, D.A., 1994. Nucleation and growth of faults in brittle rocks, J. Geophys. Res., 99, 18159–18173.

[36] Rubin, A.M., 1995. Propagation of magma filled cracks. Ann. Rev. Earth Planet. Sci. 8, 287–336.

[37] Rubin, A.M., 1993. Tensile fracture of rock at high confining pressure: Implications for dike propagation. J. Geophys. Res. 98, 15, 919–15, 935.

[38] Schock, R.N. & Louis, H., 1982. Strain behavior of a granite and a graywacke sandstone in tension. J. Geophys. Res., 87, 7817–7823.

[39] Schock, R.N., 1977. The response of rocks to large stresses. Impact and explosion cratering D.L. Roddy, R.O. Pepin, and R.B. Merrill (editors), Pergamon press, New York, NY.

[40] Scholz, C.H., Dawers, N.H., Yu, J.-Z., Anders, M.H. & Cowie, P.A., 1993. Fault growth and fault scaling laws: Preliminary results, J. Geophys. Res., 98, 21951–21961.

[41] Scholz, C.H., 1990. The mechanics of earthquakes and faulting, pp. 439, Cambridge University, England.

[42] Turcotte, D.L., Newman, W.I. & Shcherbakov, R., 2003. Micro and macroscopic models of rock fracture. Geophys. J. Int. 152, 3: 718–728.

[43] Vermilye, J.M. & Scholz, C.H., 1998. The process zone: A microstructural view of fault growth, J. Geophys. Res., 103, 12223–12236.

[44] Weinberger, R., Baer, G., Shamir, G. & Agnon, A., 1995. Deformation bands associated with dyke propagation in porous sandstone, Makhtesh Ramon, Israel, in Physics and Chemistry of Dykes, eds. Baer & Heimann, 95–112, Balkema, Rotterdam.

[45] Weinberger, R., Lyakhovsky, V., Baer, G. & Agnon, A., 2000. Damage zone around en-echelon dyke segments in porous sandstone. J. Geophys. Res. 105:3115–3133.

[46] Willemse, E.J.M. & Pollard, D.D., 1998. On the orientation and patterns of wing cracks and solution surfaces at the tips of a sliding flaw or fault. J. Geophys. Res., 103, 2427–2438.

[47] Yukutake, H., Fracturing process of granite inferred from measurements of spatial and temporal variations in velocity during triaxial deformation, J. Geophys. Res., 94, 15639–15651, 1989.

[48] Zietlow, W.K. & Labuz, J.F., 1998. Measurements of the intrinsic process zone in rock using acoustic emission". Int. J. Rock Mech. Min. Sci., 35, 291–299.

[49] Zoback, M.D. & Byerlee, J.D., 1975. The effect of microcrack dilatancy on the permeability of Westerly granite. J. Geophys. Res., 80-B, 752–755.

Earthquakes and Acoustic Emission – Carpinteri & Lacidogna (eds)
© 2007 Taylor & Francis Group, London, ISBN 978-0-415-44402-6

SPECTRAL ELEMENT SIMULATION OF RUPTURE DYNAMICS

J.-P. Vilotte & G. Festa
Department of Seismology, Institut de Physique du Globe de Paris, 75252 France

ABSTRACT

Numerical modelling is an important tool, allowing to have access to the different time and space scales carried by the nucleation and propagation of the rupture. In this paper, the spectral element method, which is widely recognized as a powerful tool in regional and global seismology, is applied to the study of the faulting process. Specifically, the importance of the nucleation process in the supershear behavior of inplane cracks is detailed.

1 INTRODUCTION

Earthquake source dynamics is one of the most important problems in geophysics, providing key elements for the prediction of strong ground motion as well as for the understanding of the physics of earthquake initiation, propagation and healing. Recent studies of rupture processes of selected earthquakes, together with the development of dense digital seismic networks in several seismogenic areas (California, Japan), have shed new lights on the seismic rupture propagation and nucleation. Specifically, they have shown evidence for supershear rupture during the 1999, Izmit, Turkey and the 2001, Kunlunshan, Tibet earthquakes, indentifying the fundamental role that friction and fault geometry play in earthquakes.

In spite of the improvement in computer facilities, rupture propagation modelling remains a difficult problem. Rupture propagation on a major earthquake fault is controlled by the properties of nonregular friction laws that describe the dissipation within the fault interface. Through the initial absolute stress along the fault, friction controls how the rupture initiates, grows from static state to sub or intersonic speeds, and finally stops. Rate- and state-dependent (Dietrich [1]) or simpler slip-weakening (Ida [2]) laws have long been advocated in seismology and both involve an intrinsic finite-length scale which play a major role in the control of the nucleation phase and of the behavior of the rupture front.

An essential requirement to study dynamic faulting is an accurate and robust method for the numerical modelling of earthquake rupturing which allows the resolution of the different time and space scales involved in the nucleation phase, the rupture front propagation and the short wave radiation owing to the heterogeneities and the geometrical complexity of the faults. Two methods have been widely used in seismological applications: the boundary integral equations method (BIEM) and the finite differences (FD). BIEM, pioneered by Das and Aki [3] can handle non planar faults, but it is limited to homogeneous or piecewise homogeneous media, at least in their current implementation. FD, introduced by Madariaga [4] for the study of seismic ruptures, and been since developped by numerous authors, can be used to study rupture propagation in heterogeneous elastic media but with simplified fault geometries.

Spectral element method (SEM) locates at the center of this framework, combining the geometrical flexibility of finite elements with the convergence rate of the spectral methods and it already demonstrated a powerful tool in modelling the seismic wave propagation at regional and global scales (Komatitcsh and Vilotte [5], Capdeville et al. [6]). Taking advantage from the variational approximation, the rupture modelling along non planar faults through friction laws can be introduced in SEM as an interface condition, in constrast with finite differences, which generally deal with velocity-stress staggered grids, where mixed boundary conditions require necessarily some interpolation. Hence, such techniques are expected to require a larger number of points to provide the same spatial resolution as SEM.

2 RUPTURE MODELLING WITH SEM

From a seismological point of view, the fault is a discontinuity surface of kinematic fields (displacement, velocity and acceleration) embedded in a linear elastic (or viscoelastic) medium. Using a domain decomposition representation, the elastic bulk is separated into two non overlapping regions, independent each of other, whose only interaction is established through the boundary conditions on the common interface. These relationships should account for the continuity of the traction and the frictional laws on the fault surface. On the external surface of the two different domains homogeneous boundary conditions (rigid or free surface) can be imposed instead.

Two terms contribute to the traction on the fault surface: a static contribution, which usually is defined before the rutpuring process and for this reason it is also referred to as initial stress, and the dynamic perturbation, which is generally lower in magnitude and comes from both the elastic bulk and the rupture itself, progessing elsewhere on the fault surface. The total traction, finally, constitutes the kernel of the boundary conditions on the fault plane, that can be summarized in the following.

The contact condition (or Signorini law) requires that either the two sides of the fault do not overlap, or, if opening occurs, both sides act like a free surface. Formally, this yields:

$$\delta u_N \geq 0; \; T_N \geq 0; \; \delta u_N T_N = 0; \tag{1}$$

where T is the traction, δu is the displacement discontinuity across the surface and the index N refers to the normal component of any field.

On the other hand, the frictional condition (Coulomb law) constrains the fault to be at rest until the tangential traction reaches a threshold value, proportional to the normal traction. While sliding, the ratio between tangential and normal traction is controlled by the friction. The formal statement of this condition is:

$$\Phi(\mathrm{T}_t, \mathrm{T}_N) := (-\|\mathrm{T}_t\| + \mu T_N) \geq 0; \quad \delta \mathrm{v}_t = \dot{\gamma}\frac{\mathrm{T}_t}{\|T_t\|}; \quad \dot{\gamma} \leq 0; \quad \dot{\gamma}\Phi = 0 \tag{2}$$

where δv is the velocity discontinuity, μ is the frictional coefficient and the index t refers to the fields as projected onto the plane locally tangential to the fault. Through this paper, we adopt a linear slip weakening law, which is widely used in seismological applications (Ida [2]).

Spectral element methods solve the variational statement of the elastodynamic equations by decomposing the elastic domains in hexahedral elements and by approximating the solutions spaces with polynomial subspaces of finite dimension. Integrals involved in the weak formulation are first referred to the single elements and then mapped back to the reference element, that is the cube in 3D. Finally numerical quadrature, applied on the Gauss-Lobatto-Legendre points (which are the same points as used to build up the polynomial basis), transforms the original differential equation into a linear system, whose mass matrix is diagonal. As a consequence, the time evolution of the system can be described through a predictior-multicorrector Newmark scheme, where velocity is recursively updated by using the displacement evaluated at the previous correction, without any matrix inversion. Traction term on the fault is naturally included in the spectral element formulation. Because of the variational feature of the method, the relationship between traction and velocity discontinuity is averaged on the whole fault, as it is seen from the elastic bulk. This smooth condition should be intersected punctually with the contact and friction laws as described above.

2.1 High frequency regularization

Interface conditions are imposed in a strong way on the fault and the discontinuities of the linear slip weakening law generate a signal with a continuous unbounded spectrum. While low frequencies are well propagated by the numerical grid, high frequencies are not scattered away from the fault itself. However, in finite differences grids they are dissipated because of the interpolation. In spectral element meshes instead, they generate spurious modes on the surface interacting in a non linear way with the rupture process. For this reason, a high-frequency regularization of the solutions is strongly required in SEM. This condition can be achieved by embedding the fault in a dispersive anisotropic medium, as obtained by an analytic continuation in the complex space, which has a trasparent behavior for low frequencies and an exponential decay absorption in the high frequency range. The cut-off frequency can be chosen close to the maximum one propagated by the numerical grid according to Courant rule. Attenuation is imposed only along the normal direction and it involves the first layer of elements surrounding the fault itself.

3 SUPERSHEAR BEHAVIOR OF INPLANE CRACKS

As an example, we analyze the supershear behavior of the crack as a function of the initial size of crack itself L and the strength parameter s:

$$s = \frac{\tau_u - \tau_0}{\tau_0 - \tau_f} \tag{3}$$

where τ_u, τ_f and τ_0 are respectively the strength value, the dynamic stress and the initial traction on the fault. To single out the characteristics of the supershear phase, we take into account the Kostrov's problem of an infinite planar fault embedded in a homogeneous medium, with homogeneous initial conditions and 2D in-plane strain. In this case normal dynamic reaction is zero, because of the symmetry. The dimensions of L should be scaled with respect to the half critical size of the crack, as defined by Andrews [7].

The rupture process is initiated by raising the initial traction a few percent above the frictional strength in a patch of length L, in order to allow the crack to propagate (Andrews [7]).

The most remarkable observation is that supershear occurs in two different ways, depending on both s and L parameters. For large values of L and correspondently small values of s, the size of the crack grows at the same time as the nucleation process evolves. At the center of the nucleation zone, the slip rate rises as in a 1D model, with very weak stopping phases coming back into the active zone from the crack tip. When crack speeds up to Rayleigh velocity, slip rate decreases in the central zone to the final constant value, which is close to the S wave velocity (the scaling factor is $\mu/\Delta\tau$ *which is itself unitary*). The supershear peak grows away from the crack front, without changing the shape of the reaction function (Fig. 1-a). It progresses initially at about $\sqrt{2}v_s$, but it accelerates

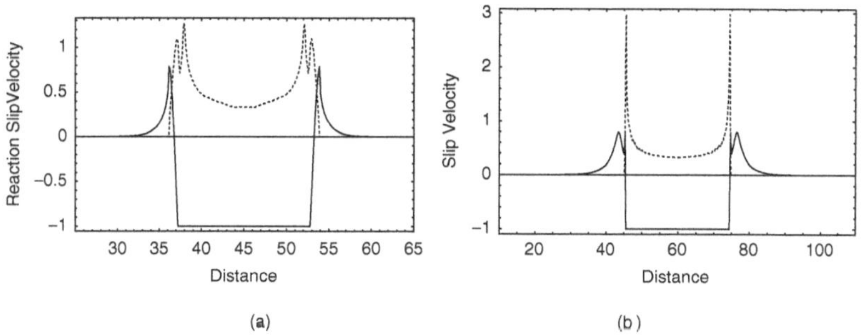

Figure 1: Slip velocity (dashed curve) and reaction (solid curve) for a supercritic crack (panel a) or a subcritic crack (panel b). In the first case supershear peak grows away from the crack front, without discontinuity. In correspondence, the cohesive zone extends (reaction function bends forward). For a subcritic crack, supershear in carried by the S wave peak in the reaction, allowing the cohesive zone to jump ahead of the front.

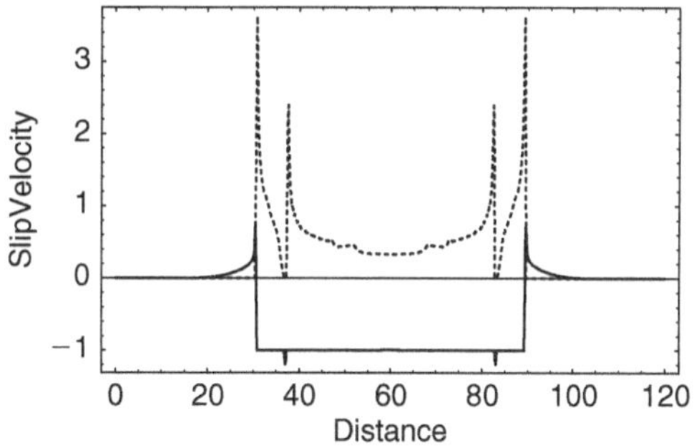

Figure 2: Slip velocity (dashed curve) and reaction (solid curve) during the supershear. While the crack propagates, external peak remains separated from the internal crack front, with a healing phase moving at the S wave velocity. It is due to a discharge in the reaction, which drops below the dynamic frictional strength.

rapidly toward P wave velocity, which is never reached, anyway. The supershear peak is always joined to the internal crack front and the peak amplitude becomes more and more important with respect to the internal peak during the propagation of the rupture. We refer this process to as a supercritic crack.

On the other hand, for large s and small L the nucleation phase progress very slowly (subcritic crack). Slip velocity at the center of the initial patch departs very early from the 1D solution, because of stopping phases coming from the boundaries of initial crack. After the nucleation phase, the amplitude decreases to a minimum, lower than the asymptotic value, because most energy is spent to allow the crack to progress. In this case, the crack propagates long time at Rayleigh speed, just to allow the S wave peak in the tangential traction to reach the threshold value (Fig. 1-b). Consequently, the cohesive zone jumps ahead of the rupture front, generating a pulse separated from the internal crack. After an initial adjustment phase, a healing phase propagating at the S wave speed occurs. The latter is carried by a discharge in the reaction, which drops below the dynamic frictional strength τ_f. In this case too, supershear peak accelerates asymptotically toward the P wave velocity, which is this problem (Burridge [8]).

3.1 Pulse domain

The importance of different signatures in the supershear phase is not only a numerical results obtained with simplified frictional conditions, but supershear has been always observed in analogical experiments with rock samples (Rosakis et al. [9]). Hence, we move to investigate in detail the supershear domain in which the pulse is present, for the same numerical experiment as described in the previous section. In this case, the threshold τ_u and the dynamic value τ_f are assumed constant and changes in the strength parameter s are implied only by the initial stress τ_0. Four different have been observed: the crack does not propagate but it stops after a while; the crack moves at the Rayleigh velocity speed; the S wave peak in the reaction reaches the frictional strength allowing the cohesive zone to jump ahead of the crack tip; the crack accelerates at supershear velocities without an advancing pulse. These domains are individuated through experimental points in Fig. (3), obtained with numerical simulations. The boundary between the crack propagation and no propagation $L_n(s)$ seems to be fit by a straight line with an almost flat slope. In fact, since τ_u and τ_f are fixed, and τ_0 is at the threshold in the nucleation zone, the available energy is the same independently of the strength parameter. If we put barriers (an infinite threshold value) at the boundaries of the nucleation zone, the signal will grow indefinitely or will stop only as a function of L. If the threshold is finite, sooner or later it will reached, and energy will be subtracted to the nucleation, in order to allow the crack to propagate. For two nearby s values, the energy necessary to the crack growth is not significantly different, and therefore, we expect a slow increase of L_n as s increases.

The boundary between pulse and no pulse zone $L_p(s)$, instead, may not be described by a straight line. In fact, with a weak frictional strength (small values of s), supershear always occurs without pulse. This limit can be estimated at about $s = 0.65$. For $s > 0.65$, a pulse occurs in the range $L_n(s) < L < L_{\mathrm{p}}(s)$. Moreover, this range remains finite for $s = 0.65$.

The graph has not many points for large s values, because the computational effort and storage requirements become prohibitive: in this range, larger and larger models should be used to capture the supershear phase. We tested for $s = 1.7$, the cases $L = 10$, $L = 20$ and $L = 40$, and we always observed a pulse carried on by the S peak. Moreover, for $s > 1$, the boundary between the two supershear domains seems to rise as an exponential function.

For $s = 1.8$ and reasonable L (just up to $L = 20$) values, the S wave peak in the reaction is seen to do not increase its amplitude with time and we are confident

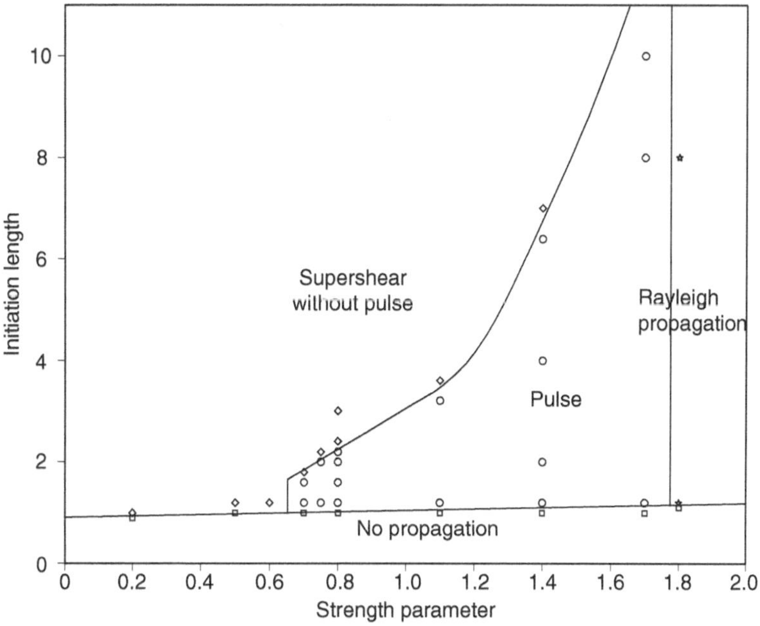

Figure 3: Behavior of crack as a function of the strength parameter s and the size of the initial patch *s*. Four different cases are possible: the crack does not propagate (squares), the crack accelerates moving at supershear speeds with a pulse in the slip velocity ahead of the crack tip (circles), the crack accelerates at supershear speeds without a pulse in advance (parallelograms) or finally the crack propagates at Rayleigh wave speed without any supershear phase (stars).

that it will not more, because of the self similarity. It agrees with the analytic limit of $s = 1.77$ discussed by Burridge [8]. Anyway this condition is obtained for a two value frictional law. For very large L ($L \rightarrow \infty$), however, the supershear may occur, since a large amount of energy is available before the cohesive zone shrinks.

References

[1] Dieterich, J.H., "Modelling of rock friction I: experimental results and constitutive equations", J. Geophys. Res., 84, 2161–2168, 1979.
[2] Ida, Y., "Cohesive force across the tip of a longitudinal-shear crack and Griffith's specific surface energy", J. Geophys. Res., 77, 3796–3805, 1972.
[3] Das, S., and Aki, K., "A numerical study of two-dimensional spontaneous rupture propagation", Geophys. J. R. Astr. Soc., 50, 643–668, 1977.
[4] Madariaga, R., "Dynamics of an expanding circular fault", Bull. Seism. Soc. Am., 66, 639–666, 1976.

[5] Komatitcsh, D., and Vilotte, J.-P., "The Spectral element method: an efficient tool to simulate the seismic response of 2D and 3D geological structures", Bull. Seism. Soc. Am., 88(2), 368–392, 1998.
[6] Capdeville, Y., Chaljub, E., Vilotte, J.-P., and Montagner, J.-P., "Coupling the spectral element method with a modal solution for elastic wave propagation in global earth models", Geophys. J. Int., 152, 34–67, 2003.
[7] Andrews, D.J., "Dynamic plane-strain shear rupture with a slip-weakening law, calculated by a boundary integral method", Bull. Seism. Soc. Am., 75(1), 1–21, 1985.
[8] Burridge, R., "Admissible speeds for plane-strain self-similar shear cracks with friction but lacking cohesion", Geophys. J. R. Astr. Soc., 35, 439–455, 1973.
[9] Rosakis, A.J., Samudrala, O., and Coker, D., "Intersonic shear crack growth along weak planes", Mat. Res. Innovat., 3, 236–246, 2000.

1.2 Mechanics of friction and faulting

Earthquakes and Acoustic Emission – Carpinteri & Lacidogna (eds)
© 2007 Taylor & Francis Group, London, ISBN 978-0-415-44402-6

NON-LINEAR SLIP-WEAKENING IN A ROTARY GOUGE FRICTION EXPERIMENT

G. Chambon[1], J. Schmittbuhl[1,3] & A. Corfdir[2]
[1] Geology Lab., Ecole Normale Supérieure, Paris, France
[2] CERMES, ENPC/LCPC, Institut Navier, Champs sur Marne, France
[3] Institut de Physique du Globe, Strasbourg, France

ABSTRACT

Two families of friction law are classically introduced to describe mechanical fault instabilities: slip weakening and rate and state friction laws. Generally opposed, we propose here, on the basis of our experimental results, to combine them in a single unified law. Using a large-displacement ring-shear apparatus for thick gouge samples (confinement: 0.1–0.5 MPa), we observe that slip-weakening largely dominates rate and state effects. The effective friction coefficient μ_{eff} is found to undergo a power-law decrease with imposed slip δ: $\mu_{eff} = \mu_0 + A\delta^{-\beta}$, with $\beta = 0.4$.

Although no characteristic length scale really exists, the main decrease of effective friction occurs over about 50 cm of slip. This appears quantitatively consistent with seismological data both in terms of typical weakening distances and characteristic rupture energies. Rate and state effects are involved over significantly smaller scales: $dc \approx 100\,\mu m$. On the micro-scale, grain attrition exists. An image correlation technique reveals that the active interface is significantly wider than suggested by the microstructures. Slip-weakening appears caused by a progressive mechanical decoupling between a shear band and the bulk of the sample. Direct normal stress measurements along the shear interface have been possible and show a negligible influence of hoop stresses. Implications for faults are finally discussed.

1 INTRODUCTION

Real faults generally consist in complex three-dimensional interfaces comprising thick layers of cataclastic gouge and damaged rocks (*e.g. Chester and Chester* [1]). In modeling studies, however, these thick structures are usually treated as perfectly thin interfaces, and their mechanical properties reduced to an effective friction law (*e.g. Scholz* [2]). The role of the friction law is to prescribe the evolution of the fault effective coefficient of friction as a function of the relevant physical parameters: slip, slip rate, asperity status, fault history, fault morphology, etc. In particular, the friction law should describe the physical mechanisms responsible for fault weakening during the initiation and development of seismic instabilities (earthquakes).

Two principal forms of friction laws coexist in the literature: the rate- and state-dependent friction (RSF) laws and the slip-weakening laws. In the RSF formulation, friction depends on the slip rate and on a set of variables characterizing the "state" of the frictional interface. This type of laws has been formulated on the base of numerous experimental results and appears to be applicable for a wide range of materials (*Marone* [3]). On the other hand, slip-weakening laws prescribe that the coefficient of friction essentially depends on slip displacement. They are less supported by experimental data than RSF laws (see nevertheless *e.g. Ohnaka and Shen* [4]), but are frequently employed for earthquake modeling owing to their relatively easy numerical implementation.

2 EXPERIMENTAL SETUP

Our experiments were performed in an annular, simple shear apparatus (Figure 1). The gouge sample is ring-shaped, with both height and width of 100 mm (inner radius R_1 mm, outer radius $R_2 = 200$ mm). Shear is applied by imposing a rigid rotation of the inner steel cylinder. As an analogy with faults, the displacement δ along sample inner boundary is called *slip*. Slip velocity $v = \dot{\delta}$ can be prescribed

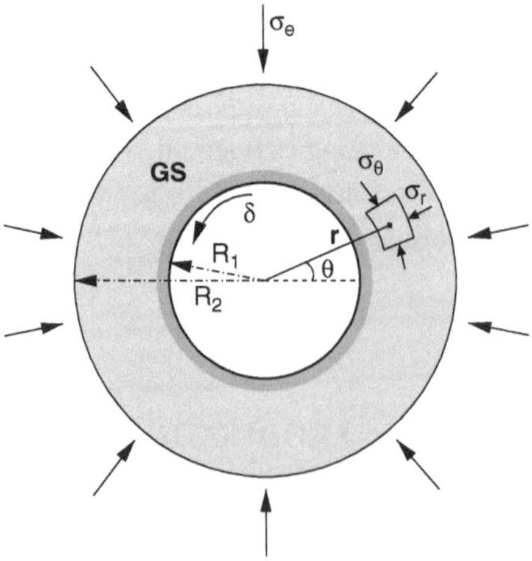

Figure 1: Sketch of the rotary setup (ACSA): The sample (GC) is sheared from the rotation of the inner cylinder of radius R_1 and confined (σ_e) at a radius R_2 owing to a water cell.

in the range 1.7–100 μm·s^{-1}. A water cell connected to a pressure-volume controller is placed around this jacket and insures the radial confinement of the sample σ_e.

Fault gouge is simulated in this study using an angular, quarry sand containing more than 99% of quartz. Working with submillimetric to millimetric particles, the thickness of our samples (perpendicular to the sliding direction) always exceeds 100 grains. This significant thickness has to be contrasted with the narrower samples used in most other gouge friction experiments (*e.g. Marone* [3]).

The torque Γ exerted by the driving system on the rotating cylinder is measured and converted in a shear stress τ. Figure 2 shows a significant weakening of shear stress τ with imposed slip. This weakening stops during (arbitrary long) hold periods at zero velocity, and proceeds as soon as shear is resumed. It is thus really slip-, and not time-, induced. We also see in Figure 2a, that the amounts of slip required for a complete saturation of the shear stress drop are surprisingly large compared to grain size, typically about 0.5 m. In Figure 2a, we show that slip-weakening is not observed only during the initial shear (IS), but also, in a very reproducible way, during sense reversal (SR) and stress drop (SD) phases.

Figure 2c shows shear stress variations when prescribed velocity changes are imposed during a shear phase. The frictional strength of our synthetic gouge

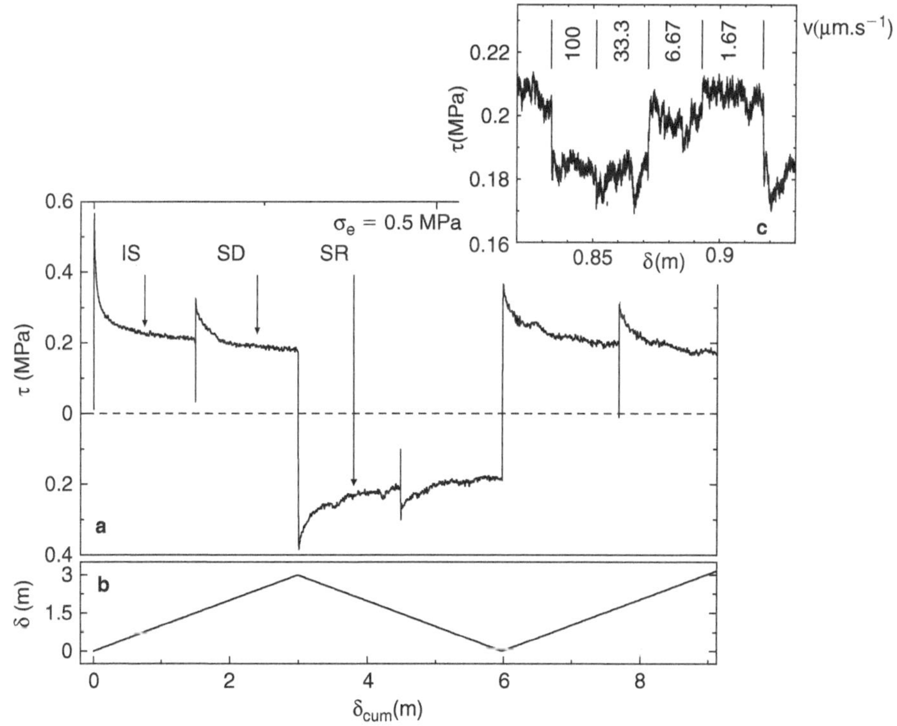

Figure 2: Response of the sample during successive shear phases: evolution of the shear stress (a), of the slip (b). Response to velocity changes is shown in (c).

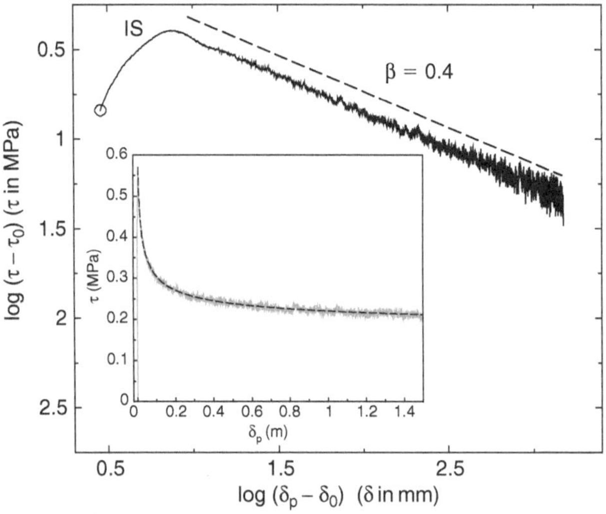

Figure 3: Power-law slip weakening during an initial phase in logarithmic scales. The inset shows the same data in a linear diagram together with the proposed fit.

samples therefore includes velocity weakening. Velocity-induced variations in shear stress, however, never exceed a few percents, and sometimes hardly emerge from the noise level. Velocity-weakening thus constitutes a second-order process compared to the major slip-weakening trend.

3 NON-LINEAR SLIP WEAKENING

Figure 3 shows that the post-peak decrease of shear stress τ with partial slip δ_p follows a linear path in log-log coordinates. This property appears particularly evident during initial shear phases: the linear decrease can then be observed over more than 2 slip decades. It holds generically, however, for all the shear phases that we studied, regardless of the preceding shear history and restrengthening events. The slip-weakening process can thus be modeled by a power law of the form:

$$\tau\left(\delta_p\right) = \tau_0 + \Delta\tau\left(\frac{\delta_*}{\delta_p - \delta_0}\right)^{\beta}, \tag{1}$$

where, remarkably, the characteristic exponent β is systematically equal to 0.4 ± 0.05.

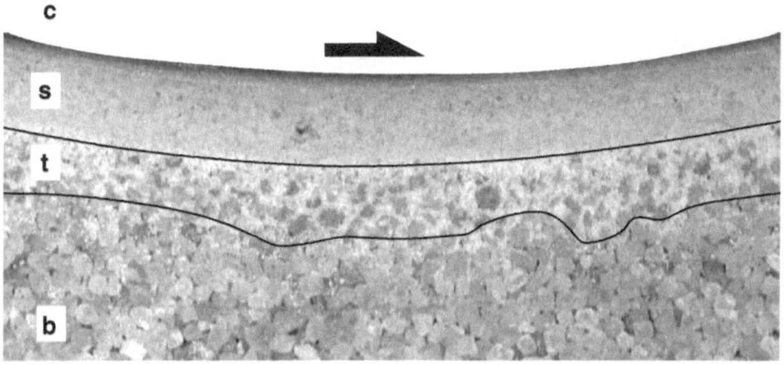

Figure 4: Micro-structures along the inner cylinder (c). A transition zone (t) separates the shear band (s) where particle flow is high and comminution intensive from the bulk (b). The scale is given by the particles which are 1 mm large.

4 MICRO-STRUCTURAL OBSERVATIONS

Two windows pierced in the apparatus enable to visualize the sample lower surface through a bottom glass plate. The observation zone is about 10 × 10 cm. A digital camera (resolution: 1536 × 1024px) has been installed below one of the windows to take series of pictures during the runs. Observations through the two windows pierced in the apparatus clearly reveal localization of deformation (Figure 4). Specifically, as soon as the peak strength is reached in a shear phase, grain displacements are found essentially confined to a narrow interfacial layer around the inner cylinder. In agreement with numerous other studies (*e.g. Mülhaus and Vardoulakis* [5]), the width of this shear band corresponds to 6–7-grain diameters. Furthermore, this width remains remarkably constant with ongoing shear, up to at least 37 m of cumulative slip.

In sand samples, strain localization appears associated with an active grain comminution process. As shown in the photo of Figure 4, imposed slip creates a fine cohesive powder which progressively fills in the porosity of the shear band. We also observe that angularities of the initial particles tend to get rounded inside the band.

In a recent paper (*Chambon* et al. [6]), we presented results when Correlation Image Velocimetry (CIV) is applied to digital pictures. While measurements inside the shear band are corrupted by grain comminution, we were able to recover, with excellent accuracy, the local strain field inside the bulk of the sample. Indeed, even once localization is established, the bulk remains submitted to slow and intermittent deformation. We identified in particular a slow relaxation process active in the 10 or 20 particle-wide annulus surrounding the shear band. We showed that this slow relaxation actually corresponds to a slow and progressive decoupling between the shear band and the bulk of the sample.

5 FAULT IMPLICATIONS

Regardless of their formulation, a recurrent concern with laboratory-derived friction laws is their quantitative discrepancy with natural, seismological data. Shear rupture energies G_c, in particular, hugely differ between friction laboratory tests ($10^{-2} - 10^{0}\,\mathrm{J{\cdot}m^{-2}}$) and real earthquakes ($10^{6} - 10^{8}\,\mathrm{J{\cdot}m^{-2}}$) (*Ohnaka* [7]). Furthermore, this rupture energy is generally treated as a material parameter by classical friction laws, whereas various seismological results suggest that it does actually depend on earthquake size (*Ide and Beroza* [8]). In the same spirit, the characteristic slip scales involved in laboratory weakening mechanisms are about $10^{-6} - 10^{-4}\,\mathrm{m}$, while earthquake inversions usually yield weakening distances D_c in the range$10^{-2} - 10^{0}\,\mathrm{m}$ (*Guatteri and Spudich* [9]).

In seismological studies, the slip-weakening processes active on faults during earthquakes are usually quantified in terms of two linked parameters: the fracture energy G_c and the characteristic weakening displacement D_c. To examine whether our laboratory results could be extrapolated to real faults, we computed the parameters G_c and D_c associated to the dominant slip-weakening process of our experiments.

The fracture energy G_c is computed for each shear phase by integrating the τ versus δ_p relationship (*Rice* [10], *Ohnaka* [7]):

$$G_c = \int_{\delta_i}^{\delta_f} \left[\tau(\delta) - \tau(\delta_f)\right] d\delta. \tag{2}$$

Where δ_f represents the partial slip at the end of the considered shear phase and δ_i is defined by: $\tau(\delta_i) = \tau(\delta_f)$ but before the peak stress. Typically we obtained $G_c \approx 2 \times 10^{4}\,\mathrm{J{\cdot}m^{-2}}$ for $\sigma_e = 0.5\,\mathrm{MPa}$. An essential feature of the power law slip-weakening exhibited in our experiments, is that G_c does not constitute intrinsic material parameters. Integrating expression (1), one finds the following scaling relationship between the fracture energy G_c and the quantity of slip s undergone by the sample during a shear phase:

$$G_c \sim s^{\alpha}, \tag{3}$$

where $\alpha = 1-\beta = 0.6$.

From a wide data compilation of seismological data, *Ide and Beroza* [8] showed that the energy radiated by earthquakes is roughly proportional to the seismic moment M_0 over more than 14 orders of magnitude in M_0. This observation, together with the classical result that earthquake stress drop is independent of M_0, indicates that the fracture energy should be proportional to earthquake slip: $G_c \sim s$. In an independent study, *Abercrombie and Rice* [11] directly evaluate the fracture energy of various earthquakes and end up with a

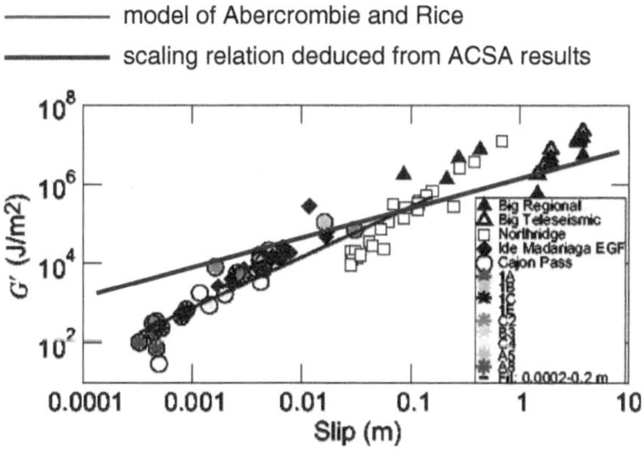

Figure 5: Scaling of the fracture energy after *Aberbercrombie and Rice* [11]. Superimposed is the fit from our scaling: $G_c \sim s^{0.6}$.

similar, though slightly different, scaling relationship: $G_c \sim s^{1.3}$ (see Figure 5). A scaling exponent of 0.6, as predicted from our experiments, would however be compatible with the data presented by *Abercrombie and Rice* [11] (see Figure 5).

References

[1] F. M. Chester and J. S. Chester, "Ultracataclasite structure and friction processes of the Punchbowl fault, San Andreas system, California," *Tectonophysics*, vol. 295, pp. 199–221, 1998.

[2] C. H. Scholz, *The Mechanics of Earthquakes and Faulting*. 1997.

[3] C. Marone, "Laboratory-derived friction laws and their application to seismic faulting," *Annu. Rev. Eath Planet. Sci.*, vol. 26, pp. 643–696, 1998.

[4] M. Ohnaka and L.-F. Shen, "Scaling of the shear rupture process from nucleation to dynamic propagation: Implications of geometric irregularity on the rupturing surfaces," *J. Geophys. Res.*, vol. 104, pp. 817–844, 1999.

[5] H. B. Mühlhaus and I. Vardoulakis, "The thickness of shear bands in granular materials," *Géotechnique*, vol. 35, no. 3, pp. 271–283, 1987.

[6] G. Chambon, J. Schmittbuhl, A. Corfdir, J.-P. Vilotte, and S. Roux, "Shear with comminution of a granular material: Microscopic deformations outside the shear band," *Phys. Rev. E*, vol. 68, 2003.

[7] M. Ohnaka, "A constitutive scaling law and a unified comprehension for frictional slip failure, shear fracture of intact rock, and earthquake rupture," *J. Geophys. Res.*, vol. 108, no. B2, 2003.

[8] S. Ide and G. C. Beroza, "Does apparent stress vary with earthquake size?," *Geophys. Res. Lett.*, vol. 28, no. 17, pp. 3349–3352, 2001.

[9] M. Guatteri and P. Spudich, "What can strong–motion data tell us about slip–weakening fault– friction laws?," *Bull. Seismol. Soc. Am.*, vol. 90, pp. 98–116, 2000.

[10] J. R. Rice, "The mechanics of earthquake rupture," in *Physics of the Earth's Interior* (A. M. Dziewonski and E. Boschi, eds.), pp. 555–649, Italian Physical Society/North Holland Publ. Co., 1980.

[11] R. E. Abercrombie and J. R. Rice, "Can observations of earthquake scaling constrain slip weakening?," *Geophys. J. Int.*, 2003. submitted.

Earthquakes and Acoustic Emission – Carpinteri & Lacidogna (eds)
© 2007 Taylor & Francis Group, London, ISBN 978-0-415-44402-6

DOES ELASTIC REBOUND THEORY APPLY TO SEISMIC FAULTS?

A. Ziv[1], A. Cochard[2] & J. Schmittbuhl[3,4]
[1]BenGurion University of the Negev, BeerSheva, Israel
[2]EES Dept, Ludwig-Maximilians-Universität, Munich, Germany
[3]Geology Lab., Ecole Normale Supérieure, Paris, France
[4]Institut de Physique du Globe, Strasbourg, France

ABSTRACT

We examine the evolution of and the exchange between two forms of elastic energies stored in the quasi-dynamic fault model of Ziv and Cochard [1]. The first, E_{tect}, is due to the integrated slip deficit accumulated between the plate boundaries and the fault surface, and the second, E_{fault}, is the elastic energy stored as a result of differential slip along the fault surface. Time series of these energies show that large earthquakes occur during the descending portions of the E_{tect}-curve, and close to the maxima of the E_{fault}-curve. Interestingly, these results are not in agreement with the classical view of the elastic rebound theory (Reid, [2]). While the elastic rebound theory predicts that large earthquakes occur at the maxima of the E_{tect}-curve, in our model they occur at the maxima of the E_{fault}-curve. The latter, arising from the slip heterogeneity along the fault, is not at all accounted for in the elastic rebound theory, on which present earthquake prediction models rely.

1 INTRODUCTION

According to the elastic rebound theory (Reid's, [2]) the seismic cycle consists of two phases. The first is the interseismic stage, during which elastic strain accumulates slowly with time due to the relative motion of the adjacent plates. The second is a seismic phase, during which the elastic strain that is stored in the medium is released abruptly. While the interseismic stage may last many years, the duration of the seismic phase is a few seconds or tens of seconds. This concept implies that major earthquakes occur when the elastic strain reaches local maxima, and that the timing and/or the size of major earthquakes are predictable. Indeed, present earthquake prediction models and hazard assessment rely on this theory. Nearly a century since the elastic rebound theory had been postulated, it is constructive to re-examine its underlying assumptions and to test its applicability to seismic faults.

In this study we test the validity of the elastic rebound theory in the quasi-dynamic fault model of Ziv and Cochard [1], which is a 3-D discrete model that employs a rate-and-state friction. Clearly the results of such an exercise are model dependent, and stress histories consistent with the elastic rebound theory may be observed in some models. Such is the case with the spring-slider systems (e.g., Schmittbuhl et al., [3]) and 2-D fault models (e.g., Ben-Zion et al., [4]), where occasional system-size earthquakes break periodically or quasi-periodically. The situation that is studied here is different in that the occurrence of large earthquakes is non-periodic, and the largest events do not rupture the entire model. Moreover, the distribution of event sizes is close to a power-law, with Omori type of clustering prior to and following large earthquakes.

2 THE MODEL

We model a long, vertical, strike-slip fault that is embedded in an elastic half-space (Figure 1). Similar to Rice [5] we calculate quasi-dynamic slip in a region that extends down to N_{depth}. This part of the model is represented by a computational grid of square cells that is periodic along strike, with repeating length of N_{length}. Below N_{depth} the fault slips steadily at the plate velocity, V_{plate}. In addition to being loaded by steady creep from below, the fault is subjected to an additional constant stressing rate due to displacement applied at V_{plate} rate on parallel planes located at distance W on either side of the fault plane. The horizontal and vertical dimensions of the computational grid are $N_{length} = 128$ cells and $N_{depth} = 64$ cells, respectively.

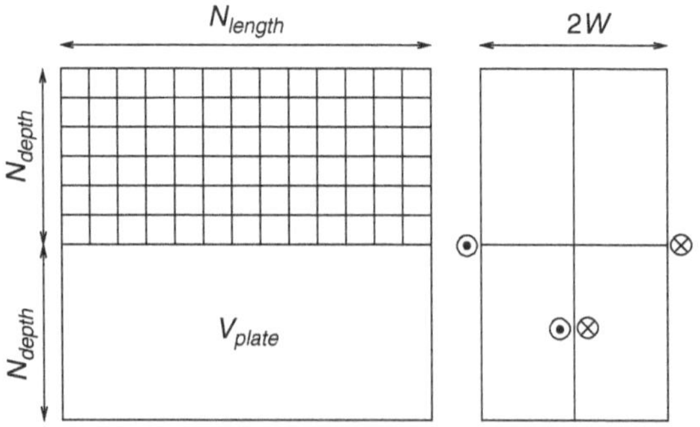

Figure 1: Schematic diagram showing strike-perpendicular (left) and strike-parallel (right) views of the model. The region over which motion is calculated is covered by $N_{length} \times N_{depth}$ computational cells. Displacement at rate V_{plate} is imposed on a co-planar substrate that extends below N_{depth} and down to $2N_{depth}$, and on fault-parallel planes located at distance W on either side of the fault plane.

Fault friction evolves with sliding speed, V, and fault state, θ, according to (Dieterich, [6]; Ruina, [7]):

$$\tau_i = \sigma_i(\mu_i^* + A_i \ln \frac{V_i}{V_{plate}} + B_i \ln \frac{V_{plate}\theta_i}{D_c}), \tag{1}$$

where the subscript i denotes the index of the computational cell, σ is the effective normal stress, μ * is the friction coefficient when the fault slides steadily at the plate velocity V_{plate}, A and B are unitless parameters, and D_c is a characteristic distance for the evolution of the state from one steady state to another. In general, the three constitutive parameters A, B and D_c may be position-dependent. Here, however, A and B vary only with depth and D_c is fixed. The state variable evolves with time, t, and slip history according to (Ruina, [8]):

$$\frac{d\theta_i}{dt} = 1 - \frac{V_i\theta_i}{D_c}. \tag{2}$$

At high slip speed, the second term on the right-hand side of (2) dominates, the state decreases exponentially with slip, and the cell weakens rapidly (Dieterich, [9]). At low speed, the state increases almost linearly with time, and the cell undergoes strengthening.

The evolution of the shear stress on cell i is written as a sum of four terms (Ziv and Rubin [11]):

$$\tau_i = \tau_i^0 + \frac{G}{W}(V_{plate}\, t - \delta_i) + \sum_i g_{ij}\delta_j - \frac{G}{2\beta}V_i. \tag{3}$$

The first term, τ_i^0, is the initial stress. The second term represents the driving stresses imparted on the fault surface due to mismatch between the total displacement on the cell in question, δ_i, and the cumulative tectonic slip imposed at rate V_{plate} at distance W on either side of the fault plane, with G being the shear modulus. The third term adds the elastic stress changes imposed on cell i due to slip on cells j, with g_{ij} being a 3-D elastic kernel. While cells labeled with i extends from the free surface down to N_{depth}, the summation over j extends from $2 \times N_{depth}$ below the free surface to $2 \times N_{depth}$ above the free surface. Finally, the forth term is an inertial term that embodies the quasi-dynamic approximation of Rice [5]. The factor $G/2\beta$, with β being the shear wave speed, is often referred to as the "radiation damping".

Stress balance yields (after derivation with respect to time):

$$\frac{dV_i}{dt} = \left[\frac{G}{W}(V_{plate} - V_i) + \sum_i g_{ij}V_j - \frac{\sigma_i B_i}{\theta_i}\left(1 - \frac{V_i\theta_i}{D_c}\right)\right] \times \left(\frac{G}{2\beta} + \frac{\sigma_i A_i}{V_i}\right)^{-1}. \tag{4}$$

Notice that the evolution of V and θ is fully described by Equations (2) and (4). This set of differential equations is solved simultaneously at successive time steps using a Runge-Kutta algorithm.

3 ENERGY PARTITIONING

Following Ziv and Schmittbuhl [10], we introduce two elastic energy densities per unit area in the system immediately after an earthquake k: E^k_{tect} and E^k_{fault}. The first is due to the slip deficit accumulated between the plate boundaries and the fault surface, and is defined as:

$$E^k_{tect} = \frac{1}{N}\frac{G}{W}\sum_i \left[U_{tect}(t_k) - U_i(t_k)\right]^2, \qquad (5)$$

where G is the shear modulus, N is the number of computational cells, U_{tect} is the cumulative tectonic displacement increasing linearly with time, U_i is the cumulative slip on cell i increasing coseismically, and t_k is time immediately after event k. The second energy is the result of stress transfer due to slip on the fault, and is defined as:

$$E^k_{fault} = \frac{1}{2N}\sum_i \sum_j g_{ij}\left[U_i(t_k) - U_j(t_k)\right]^2. \qquad (6)$$

While E_{tect} is a measure of the mismatch between total slip of the plate boundary and cumulative slip on the fault, E_{fault} is a measure of the heterogeneity of the slip distribution.

4 RESULTS

Time series of these energies show that large earthquakes occur during the descending portions of the E_{tect}-curve, and close to the maxima of the E_{fault}-curve. On a E_{fault}-versus-E_{tect} plot, the seismic cycle has a roughly triangular shape with large earthquakes occurring at the top corner of the triangle, and the foreshocks and the aftershocks occupying the right side and left side, respectively. While both foreshocks and aftershocks dissipate tectonic energies, the cumulative effect of the foreshocks is to increase the fault disorder and the cumulative effect of the aftershocks is to reduce it.

Interestingly, these results are not in agreement with the classical view of the elastic rebound theory (Reid, [2]). While the elastic rebound theory predicts that large earthquakes occur at the maxima of the E_{tect}-curve, in our model they occur at the maxima of the E_{fault}-curve. The latter, arising from the slip heterogeneity along the fault, is not at all accounted for in the elastic rebound theory. Because present earthquake prediction models rely on the elastic rebound theory, the implications of this study for hazard assessment are evident.

References

[1] Ziv, A., and A. Cochard, Seismicity on a fault with depth-variable rate- and state- dependent friction: quasi-dynamic inherently-discrete model, in preparation.

[2] Reid, H. F., The mechanism of the earthquake, In The California Earthquake of April 18, 1906, Report of the State Earthquake Investigation Commission, Vol. 2, Washington, DC, Carnegie Institution, pp. 1–192, 1910.

[3] Schmittbuhl, J., J.-P. Villote, and S. Roux, A dissipation-based analysis of an earthquake fault model, J. Geopys. Res., 101, 27741–27764, 1996.

[4] Ben-Zion, Y., M. Eneva, and Y. Liu, Large earthquake cycles and intermittent criticality on heterogeneous faults due to evolving stress and seismicity, J. Geopys. Res., 108, 2307, doi:10.1029/2002JB002121, 2003.

[5] Rice, J. R., Spatio-temporal complexity of slip on a fault, J. Geophys. Res., 98, 9885-9907, 1993.

[6] Dieterich, J. H., Modeling of rock friction, 1. Experimental results and constitutive equations, J. Geophys. Res., 84, 2161–2168, 1979.

[7] Ruina, A., Slip instability and state variable friction laws, J. Geopys. Res., 88, 10359–10370, 1983.

[8] Ruina, A. L., Friction laws and instabilities: A quasistatic analysis of some dry frictional behavior, Ph.D. thesis, Brown Univ., Providence, R. I., 1980.

[9] Dieterich, J. H., Earthquake nucleation on faults with rate- and state-dependent strength, Tectonophysics, 211, 115–134, 1992.

[10] Ziv, A., and J. Schmittbuhl, The seismic cycle and the difference between foreshocks and aftershocks in a mechanical fault model, Geophys. Res. Lett., 30(24) 2237, doi:10.1029/2003GL018665, 2003.

[11] Ziv, A., and A. M. Rubin, Implications of rate-and-state friction for properties of aftershock sequence: quasi-static inherently discrete simulations, J. Geopys. Res., 108, 2051, doi:10.1029/2001JB001219, 2003.

Earthquakes and Acoustic Emission – Carpinteri & Lacidogna (eds)
© 2007 Taylor & Francis Group, London, ISBN 978-0-415-44402-6

CORRELATIONS OF STRESS DISTRIBUTIONS ALONG THE FAULT: FROM LABORATORY FRACTURE ROUGHNESS TO FAULT ASPERITY SQUEEZE

J. Schmittbuhl[1,4], G. Chambon [1,2], A. Hansen[2] & M. Bouchon[3]
[1]Geology Lab., Ecole Normale Supèrieure, Paris, France
[2]Physics dept, NTNU, Trondheim, Norway
[3]LGIT, J. Fourier Univ., Grenoble, France
[4]Institut de Physique du Globe, Strasbourg, France

ABSTRACT

We analyze the spatial correlations of the absolute stress field along the Nojima fault, Japan in terms of scaling invariance (self-affine scaling). Despite the small range of resolution, we obtain a signature of correlations consistent with an elastic compression of the fault roughness. Indeed, we compare the wavelet spectrum of the measured stress field to that obtained from a numerical model that fully squeezes two thick elastic media bounded by interfaces which include spatially correlated asperities (i.e. non diluted asperities). Friction is assumed to follow a Coulomb law locally and interface roughness is described using the self-affine topography that is observed over a very wide range of scales from fractures to faults, with a Hurst exponent H. We find that the stress field is also self-affine, but with a Hurst exponent $H - 1$. Fluctuations of the normal stress are shown to be important, especially at local scales with anti-persistent correlations.

1 INTRODUCTION

Numerous recent studies have been proposed to reconstruct the slip and stress histories along the fault during large earthquakes (e.g. Bouchon [1], Olsen et al. [2], Ide and Takeo [3]). Even if inversions for the same fault show discrepancies, co-seismic slip and stress drop distributions exhibit very heterogeneous patterns. Such observations suggest that either dynamical processes roughen the stress fields or strong heterogeneities of the initial stress field exist along the fault.

Absolute stress field however is more difficult to obtain. It has been obtained in the case of the Kobe, 1995 earthquake (Bouchon [4]). It relies on the rotation of the slip vector during the earthquake (Spudich et al. [5]). The absolute initial stress field shows significant spatial fluctuations (see Fig. 1). The characteristic size of asperities is of the order of 10 km with an isotropic distribution.

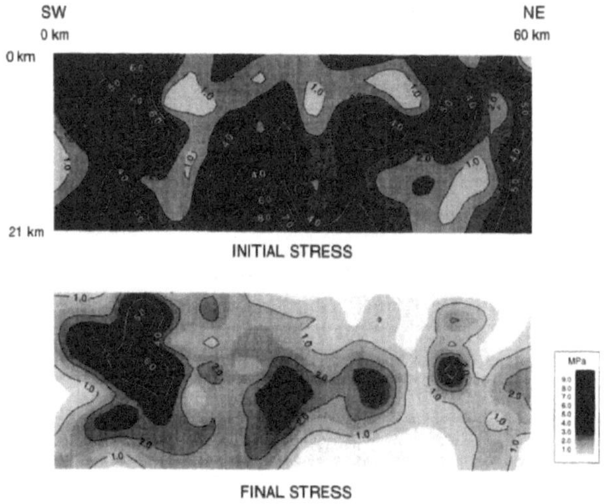

Figure 1: Absolute shear stress distributions along the Nojima Fault, Japan: before (initial stress) and after (final stress) the Kobe, 1995 Earthquake (from Bouchon et al. [4]).

Fig. 1 also shows the final stress field. It is of interest to note that although magnitude of the peaks are smaller, their positions are persistent. Accordingly, heterogeneities of the stress field are quenched along the fault and weakly sensitive to dynamical stress fluctuations due to earthquake propagation.

In this study, we analyze the possible spatial correlations of the absolute stress field using wavelet transform. We show that a power law behavior is compatible with the data and characterize the stress field fluctuations as a self-affine distribution. As a second step, we attempt to link the stress distribution along the fault to that resulting from the squeeze of fault surface asperities. For that purpose, we analyze the elastic transformation of the fault plane roughness on the basis of a boundary element model.

2 CORRELATIONS OF THE STRESS FIELD ALONG THE FAULT

We search in the initial stress field $\tau(x, z)$ presented in Fig. 1 for spatial correlations. The grid of estimated value is 61×22 with a spatial resolution of $1\,\text{km} \times 1\,\text{km}$. Actually, stress estimates are obtained from the inverted slip distribution along the fault (Bouchon et al. [4]) and an interpolation is performed at the lowest scales (<5 km). Along horizontal profiles, we search for the auto-correlation function of the local stress estimate, using wavelet transform. Because of the limited statistics, we used a sensitive tool: the Average Wavelet Coefficient Technique (Simonsen et al. [6]). It consists of an average of the wavelet transform:

$$W_{a,b} = \frac{1}{\sqrt{a}} \int_{-\infty}^{+\infty} dx \, \psi \left(\frac{x-b}{a} \right) \left| \tau(x, z = const) \right| \qquad (1)$$

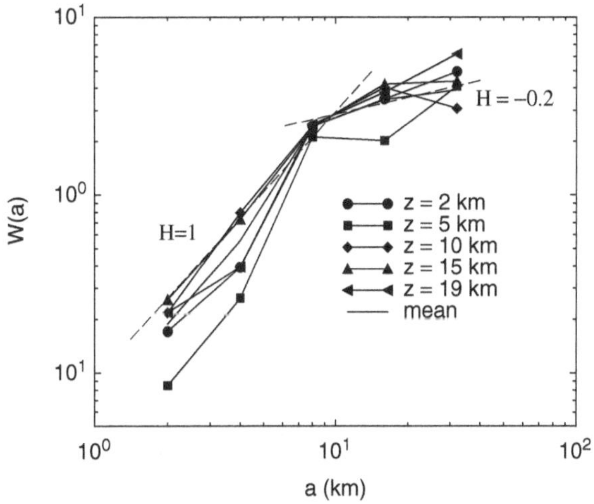

Figure 2: Average wavelet spectrum of the horizontal initial stress profiles at various depths. At small scale the spectrum is controlled by the linear interpolation procedure. At large scale, data are compatible with a power law with a slope of 0.3 in agreement with a Hurst exponent $H_\sigma = -0.2$.

where ψ is the wavelet: Daubechies wavelet of order 12. Then the wavelet coefficients are averaged over the translation factor b for each length scale a:

$$W = \left\langle W_{a,b} \right\rangle_b . \tag{2}$$

If the auto-correlation function of a stress profile scales as:

$$C(\tau(x), \tau(x+d)) \propto d^{2H_\tau} \tag{3}$$

where H_τ is the Hurst exponent of the shear stress field, then the average wavelet coefficient scales as:

$$W_a \propto a^{H_\tau + 1/2}. \tag{4}$$

Figure 2 shows the average wavelet spectra at different depths z and the mean over all accessible depths in a log-log plot. Two regimes emerge: at small scales ($2\,\mathrm{km} < a < 10\,\mathrm{km}$) a linear behavior is consistent with data interpolation. At large scales ($a > 43$ $10\,\mathrm{km}$), a power law with an exponent 0.3 ± 0.1 is a possible fit. The latter leads to a negative Hurst exponent of $H_\tau = -0.2 \pm 0.1$.

3 FAULT ROUGHNESS

Roughness of fault plane has been largely studied both at lab scales (Scholz [7]; Power and Durham [8]) and at field scale (Power et al. [9]; Schmittbuhl et al.

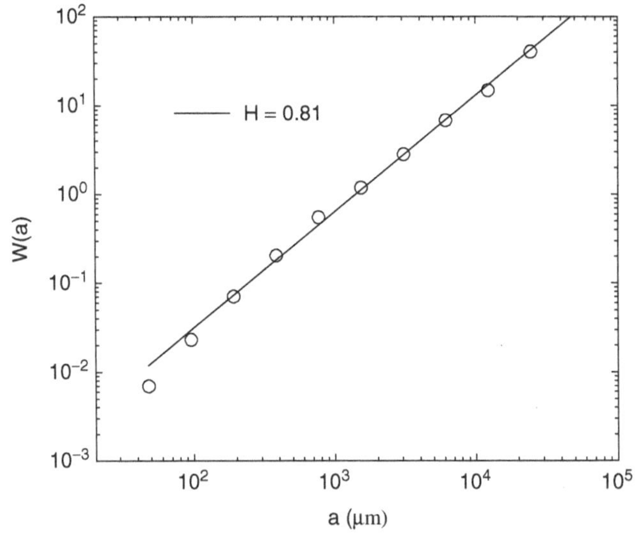

Figure 3: Average wavelet spectrum of fault roughness at laboratory scale. The sampled fault is located on the Bastille hill, Grenoble, France.

[10]). We report on a measurement of the topography of a fault surface extracted from the Bastille Hill fault near the city of Grenoble, France. A optical profiler has been used with a resolution of 3 μm for positions along the mean fracture plane and of 1 μm height estimates (Renard et al. [11]). The grid was 4100 × 873 with a mesh of 24 μm × 24 μm Analysis of the roughness has been performed using the same wavelet technique as described above, i.e. the AWC technique. Figure 3 shows the wavelet spectrum averaged along the 873 profiles.

A power law behavior with a slope of 1.31, i.e. corresponding to a Hurst exponent of $H = 0.81$, is very consistent with the data over three decades. This result is also similar to other measurements (e.g. Scholz [7]).

4 ASPERITY SQUEEZE MODEL

We attempt to link the roughness of fault surfaces to the stress field distribution owing to the analysis of the transformation of the fault asperities when submitted to a normal load. We limit ourself to an elastic deformation of the topography but include the broad range of asperity scales as observed on natural fault surfaces. The approach is based on the work of Hansen et al. [12] which consists of a boundary element modeling of the interface deformation using Fourier acceleration. The code provides an estimate of the normal stress distribution when squeezing elastically a rough surface with large range of Hurst exponents as shown in Figure 4 ($H \in [-1;2]$). We then apply the AWC technique to the produced stress field and obtained the Hurst exponent of the normal stress field

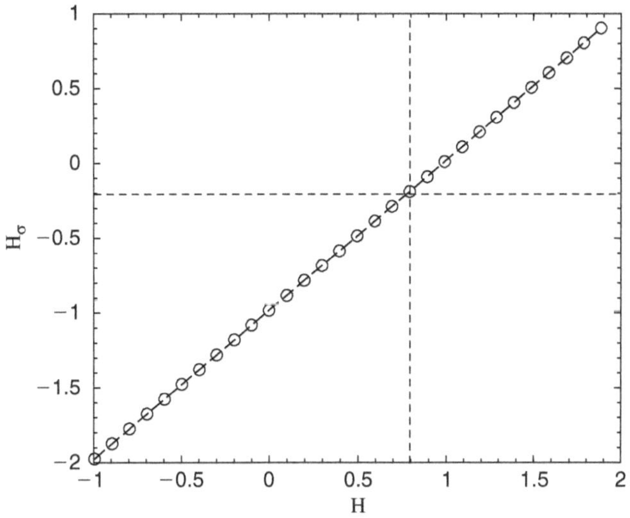

Figure 4: Estimate of the Hurst exponent H_σ of the stress field when a surface of roughness exponent H is fully squeezed in the elastic regime. The relationship: $H_\sigma = H - 1$ is remarkably fulfilled over a large range of surface geometry.

H_σ. If we assume a local Coulomb friction law (prior to the earthquake) at any point along the fault, we obtained a prediction of the Hurst exponent of the shear stress field along the fault H_τ. Therefore, if the Hurst exponent of the fault roughness is $H = 0.8$, we predict that the Hurst exponent of the shear stress field is $H_\tau = -0.2$ which is in good agreement with what is obtained at large scale (i.e. above the smoothing length scale) as shown in Figure 2.

5 CONCLUSIONS

For the 1995 Kobe earthquake, it has been possible to reconstruct not only the relative stress field during the event (stress drop), but also the absolute stress field (Bouchon et al. [4]). It appears that distributions of initial and final stresses before and after the earthquake, look very similar. This suggests that stress distribution constitutes an intrinsic property of Nojima fault, and is only slightly affected by earthquakes.

We have examined whether the dominant control of this stress distribution could be the compression of an effective fault roughness. We analyze the spatial correlations of the initial stress field in terms of scaling invariance (self-affine scaling). It can be shown (Hansen et al. [12]) that the normal stress field resulting from the full squeeze of elastically-correlated asperities characterized by a roughness exponent H is self-affine with a Hurst exponent $H - 1$. Since roughness exponents of surfaces are systematically below unity (for a wide range of scales from fractures to faults), squeeze-induced normal stresses are expected to display spatial fluctuations with anti-persistent correlations.

Initial shear stress along Nojima fault is indeed characterized by strong spatial variability. Despite a small range of resolution, its averaged wavelet spectrum is consistent with a slightly negative Hurst exponent ($H_\tau = -0.2$). A similar self-affine behavior is observed for the stress drop during the event. Total slip during the earthquake also seems to display self-affine properties, with a Hurst exponent slightly below unity ($H_u = 0.8$).

Hence, the signature of stress correlations along Nojima fault is consistent with the squeeze of a self-affine roughness. Furthermore, we also show that the initial stress field along the fault constitutes a strong guide for the development of the earthquake (stress drop and slip). Note that seismological data concern shear stresses, whereas, at the moment, squeeze models only account for normal stresses. Assuming a simple friction law, i.e. Coulomb law, we may infer from our results that stress distributions along large faults, and earthquake mechanisms, could be primarily controlled by an effective fault roughness.

References

[1] M. Bouchon, "The state of stress on some faults of the San Andreas system as inferred from near-field strong motion data," J. Geophys. Res., vol. 102, no. B6, pp. 11731–11744, 1997.

[2] K. B. Olsen, R. Madariaga, and R. J. Archuleta, "Three-dimensional dynamic simulation of the 1992 Landers earthquake," Science, vol. 278, pp. 834–838, 1997.

[3] S. Ide and M. Takeo, "Determination of constitutive relations of fault slip based on seismic wave analysis," J. Geophys. Res., vol. 102, pp. 27, 379–27,391, 1997.

[4] M. Bouchon, H. Sekiguchi, K. Irikura, and T. Iwata, "Some characteristics of the stress field of the 1995 Hyogo-ken Nanbu (Kobe) earthquake," J. Geophys. Res., vol. 103, pp. 24271–24282, 1998.

[5] P. Spudich, M. Guatteri, K. Otsuki, and J. Minagawa, "Use of fault striations and dislocation models to infer tectonic shear stress during the 1995 hyogo-ken nanbu (kobe) earthquake," B.S.S.A., vol. 88, no. 2, pp. 413–427, 1998.

[6] I. Simonsen, A. Hansen, and O. M. Nes, "Using wavelet transforms for hurst exponent determination," Phys. Rev. E, vol. 58, p. 2779, 1998.

[7] C. H. Scholz, The Mechanics of Earthquakes and Faulting. 1997.

[8] W. L. Power and W. B. Durham, "Topography of natural and artificial fractures in granite rocks: Implications for studies of rock friction and fluid migration," Int. J. Rock Mech. Min. Sci., vol. 34, pp. 979–989, 1997.

[9] W. L. Power, T. E. Tullis, S. R. Brown, G. N. Boitnott, and C. H. Scholz, "Roughness of natural fault surfaces," Geophys. Res. Lett., vol. 14, pp. 29–32, 1987.

[10] J. Schmittbuhl, S. Gentier, and S. Roux, "Field measurements of the roughness of fault surfaces," Geophys. Res. Lett., vol. 20, pp. 639–641, 1993.

[11] F. Renard, J. Schmittbuhl, J. P. Gratier, P. Meakin, and E. Merino, "Three-dimensional roughness of stylolites in limestones," J. Geophys. Res., vol. 109, no. B3, p. B03209, 2004.

[12] A. Hansen, J. Schmittbuhl, G. Batrouni, and F. D. Oliveira, "Normal stress distribution of rough surfaces in contact," Geophys. Res. Lett., vol. 27, pp. 3639–3643, 2000.

Earthquakes and Acoustic Emission – Carpinteri & Lacidogna (eds)
© 2007 Taylor & Francis Group, London, ISBN 978-0-415-44402-6

DETACHMENT WAVES AND THE ONSET OF FRICTIONAL SLIP

S. M. Rubinstein, G. Cohen & J. Fineberg
The Racah Institute of Physics, The Hebrew University of Jerusalem,
Givat Ram, Jerusalem Israel

ABSTRACT

We perform real-time measurements of the net contact area between two blocks of like material at the onset of frictional slip. We show that the process of interface detachment, which immediately precedes the inception of frictional sliding, is governed by three different types of detachment fronts. These crack-like detachment fronts differ by both their propagation velocities and by the amount of net contact surface reduction caused by their passage. The most rapid fronts propagate at intersonic velocities but generate a negligible reduction in contact area across the interface. Sub-Rayleigh fronts are crack-like modes which propagate at velocities up to the Rayleigh wave speed, V_R, and give rise to an approximate 10% reduction in net contact area. The most efficient contact area reduction ($\sim$20%) is precipitated by the passage of "slow detachment fronts". These fronts, which have not previously been observed, propagate at velocities over an order of magnitude lower than V_R. They are generated, in conjunction with intersonic fronts, by the sudden arrest of sub-Rayleigh fronts. *No* overall sliding occurs until either of the slower two fronts traverses the *entire* interface, and motion of the leading edge of the interface is initiated. Slip at the trailing edge of the interface accompanies the motion of both the slow and sub-Rayleigh fronts. We might expect these modes to be important in both fault nucleation and earthquake dynamics.

1 INTRODUCTION

The dynamics of frictional slip have been studied for hundreds of years, yet many aspects of these everyday processes are not understood. One such process is the onset of slip. First described by Coulomb and Amontons as the transition from static to dynamic friction, the onset of frictional slip is central to fields as diverse as physics, tribology, the study of earthquakes and fracture. The dynamics of friction have, traditionally, been considered to be governed by processes that occur at slow time scales, when the entire slider is in motion. Here we study the detailed processes that occur at the interface between two blocks of like material, immediately prior to the onset of frictional sliding. These processes, in which extremely short time scales are dominant, take place via 3 different types of detachment fronts which propagate along the interface. These rapid processes are fundamental to fault nucleation and the onset of slip in earthquake dynamics.

Our experiments describe the dynamics of two brittle acrylic (PMMA) blocks separated by a rough interface (each surface was roughened to an approximate roughness of 1 μm rms), consisting of many discrete randomly distributed micro-contacts. Slip is initiated by slowly increasing applied shear force, F_S, while loading the blocks with a constant normal force, F_N. The dynamics of the interface, described below, are measured by means of real-time visualization of the net contact area formed by the rough interface. The experimental apparatus, schematically shown in Figure 1a, measures the net contact area between the two rough surfaces that form the interface between the two blocks. The interface is illuminated by a laser sheet whose angle of incidence is much larger than the critical angle needed for total internal reflection from the acrylic-air interface. For the experimental conditions used, light will only effectively traverse the interface at points of contact. Thus, the intensity, I, at each point along the interface is proportional to the net contact area at that spatial

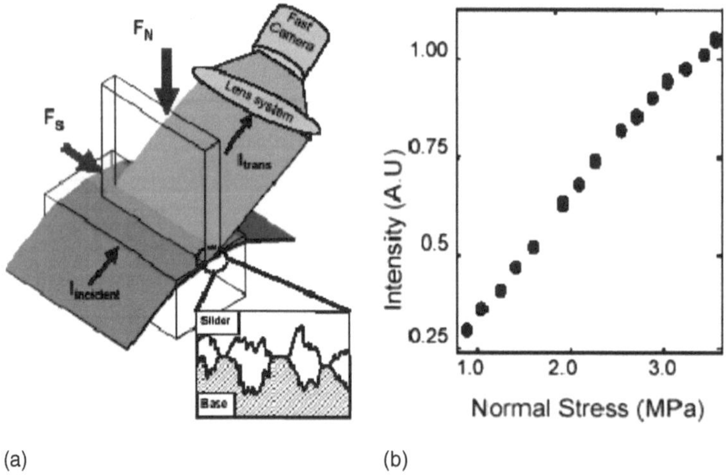

Figure 1: (a) A schematic view of the experimental apparatus, in which a laser sheet is incident on a rough interface (inset) at an angle beyond the angle for total internal reflection from a PMMA-air interface. Light can only traverse the interface at the actual points of contact. The transmitted light, imaged on a fast camera, is therefore proportional to the net contact area at each point. (b) The integrated intensity of light, hence the total net contact area, is proportional to the applied normal stress, F_N as predicted in the Bowden and Tabor theory of friction.

location. The intensity as a function of space is then imaged by a fast camera, which enables us to visualize the instantaneous net contact area along the entire interface at rates of up to 4 μsec per frame. As Figure 1b shows, the spatially integrated transmitted intensity, hence the total net contact area along the interface, is a linear function of the applied normal stress, providing a direct validation of the Bowden and Tabor picture of friction (Bowden [1], Persson [2]).

2 RESULTS

The onset of slip is immediately preceded by the interplay of three different types of coherent crack-like detachment fronts, as shown in Figure 2a and Figure 2b. Detachment is initiated by the first of these detachment fronts, the "sub-Rayleigh" fronts, which propagate at rapid subsonic velocities up to and including the Rayleigh wave speed, V_R. Sub-Rayleigh fronts lead to an approximate 10% decrease in the local contact area, immediately upon their passage. These fronts can be identified with earthquakes that propagate in the 0.2–0.8 V_R range. Upon detachment initiation, these fronts rapidly accelerate until arriving at velocities in the vicinity of V_R ($V_R = 935$ m/s in PMMA). At approximately V_R, these sub-Rayleigh fronts arrest and are *replaced* by two additional types of fronts, which are simultaneously emitted.

The more rapid of these fronts are intersonic fronts, which propagate at speeds which are considerably higher than the shear wave velocity. These types of fronts have been observed in recent experiments (Rosakis [5], Xia [6]) and calculations (Gao [3], Needleman [4]) of explosively induced shear fracture. There is also evidence for such fronts in the seismic records (Bouchon [7]) of recent earthquakes (e.g. Izmit, 1999). We will show that, although these waves traverse the entire length of the interface, they give rise to minimal (1–2%) detachment of the net contact area and negligible slip.

The second type of front which is emitted upon arrest of the sub-Rayleigh fronts is a new type of front, a "slow detachment front", which propagates an order of magnitude more *slowly* than the sub-Rayleigh fronts. Our measurements suggest that slow detachment fronts nearly always occur; either as isolated processes or in conjunction with the more rapid modes. These fronts are the dominant mechanism for interface detachment, leading to about a 20% reduction in contact area upon their passage. The slow fronts also cause a significant amount of slip while in motion, but their acoustic signatures are significantly weaker than those of the faster modes. Recent observations of slow or "silent" earthquakes (Crescentini [8], Linde [9], Rogers [10]), in which significant slip was observed with a minimal acoustic signature, may therefore be analogous to these fronts.

No overall sliding of the two blocks occurs until either of the slower two modes traverses the *entire* interface. The onset of sliding, which is initiated by the detachment fronts described above, is the transition from static to dynamic

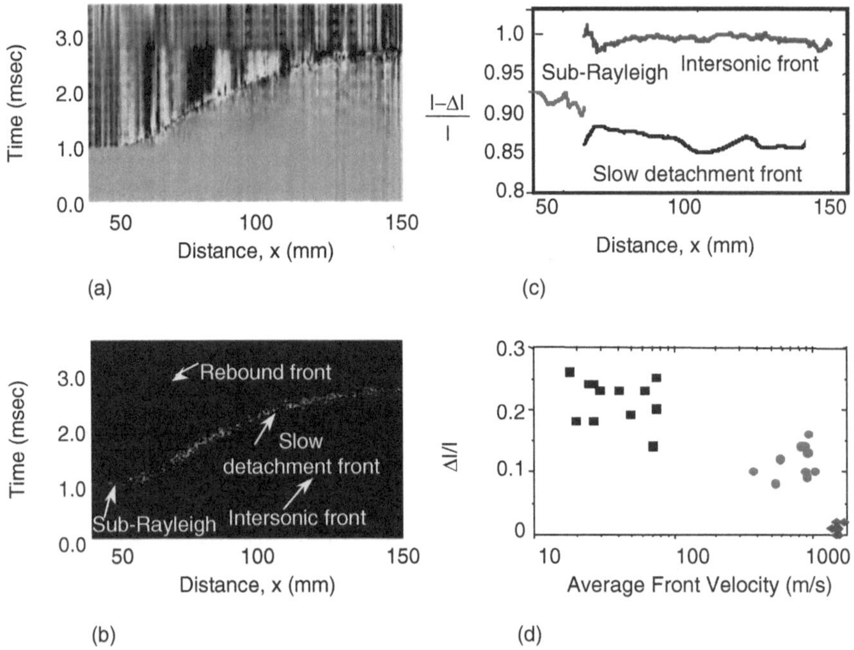

Figure 2: The dynamics of slip, *prior to overall sliding*, take place via the interplay between 4 different types of coherent crack-like fronts. (a) The transmitted intensity I(x,t), averaged in the y direction, of a typical experiment as a function of time. Successive lines spanning the 150 mm long interface were acquired at 10 μsec intervals. These intensity measurements, normalized by their initial values at each spatial point, are color-coded to reflect the change in the contact area at each spatial point as a function of time. Hot (cold) colors reflect increased (decreased) net contact area. (b) The temporal derivative, dI/dt, of the data presented in (a), color coded as above. The 4 different types are labeled within the plot. The visibility of the fronts, which can also be seen in (a), is enhanced by the derivation. (c) The relative drop in contact area, as indicated by the intensity drop (I-ΔI)/I across each of the three forward-propagating fronts shown in (a). (d) The contact area reduction, Δ I/I, in a number of different experiments for slow detachment fronts (squares), sub-Rayleigh fronts (circles) and intersonic fronts (diamonds) as a function of their measured propagation velocities. Note both the large differences in velocities and significant differences in the contact area reductions of the different fronts.

friction. The 10–20% reduction in the static verses dynamic coefficients of friction for PMMA is consistent with the overall 12% average reduction in the net contact area that we observe prior to the onset of sliding. Thus, this work elucidates both the mechanisms responsible for the transition from static to dynamic friction as well as possible key processes that take place in dynamics of fault nucleation.

References

[1] Bouchon, M., M.P. Bouin, H. Karabulut, M.N. Toksoz, M. Dietrich, and A.J. Rosakis, How fast is rupture during an earthquake? New insights from the 1999 Turkey earthquakes, Geophys. Res. Lett., 28 (14), 2723–2726, 2001.
[2] Bowden, F.P., and D. Tabor, The Friction and Lubrication of Solids, Oxford Univ. Press, New York, 2001.
[3] Gao, H.J., Y.G. Huang, and F.F. Abraham, Continuum and atomistic studies of intersonic crack propagation, J. Mech. and Phys. of Sol., 49 (9), 2113–2132, 2001.
[4] Needleman, A., An analysis of intersonic crack growth under shear loading, J. Appl. Mech., 66 (4), 847–857, 1999.
[5] Persson, B.N.J., Sliding Friction Physical Principles and Applications, Springer-Verlag, New York, 2000.
[6] Rosakis, A.J., O. Samudrala, and D. Coker, Cracks faster than the shear wave speed, Science, 284 (5418), 1337–1340, 1999.
[7] Xia, K., A.J. Rosakis, and H. Kanamori, Laboratory Earthquakes: The Sub-Raleigh-to-Supershear Rupture Transition, Science, 303 (19), 1859–1861, 2004.
[8] Crescentini, L., A. Amoruso, and R. Scarpa, Constraints on slow earthquake dynamics from a swarm in central Italy, Science, 286, 2132–2134, (1999).
[9] Linde, A.T., and I.S. Sacks, Slow earthquakes and great earthquakes along the Nankai trough, Earth and Planetary Science Letters, 203 (1), 265–275, 2002.
[10] Rogers, G., and H. Dragert, Episodic tremor and slip on the Cascadia subduction zone: The chatter of silent slip, Science, 300 (5627), 1942–1943, 2003.

Earthquakes and Acoustic Emission – Carpinteri & Lacidogna (eds)
© 2007 Taylor & Francis Group, London, ISBN 978-0-415-44402-6

CRACK-LIKE AND PULSE-LIKE DYNAMIC FRICTIONAL SLIDING

D. Coker[1], G. Lykotrafitis[2], A. Needleman[1] & A.J. Rosakis[2]
[1]Division of Engineering, Brown University, USA
[2]Division of Engineering and Applied Physics, California Institute of Technology, USA

ABSTRACT

Numerical and experimental investigation of frictional sliding under dynamic loading conditions is discussed. The configuration analyzed consists of two plates of Homalite (an elastic birefringent polymer material) connected along a planar interface. The plates are characterized as isotropic elastic materials and the interface is characterized by a rate- and state-dependent frictional law that also accounts for dependence on normal stress variations. The calculations are carried out within a framework where two constitutive relations are used: a volumetric constitutive relation between stress and strain and a surface constitutive relation that characterizes the frictional behavior of an interface. The propagation speeds of the sliding tip are found to be of the order of the longitudinal wave speed. Frictional sliding is found to occur in modes that involve uniform sliding behind the rupture front, an isolated slip pulse, multiple slip pulses or a combination of these modes. The dependence of the sliding mode on the initial compressive stress, the impact velocity and the friction parameters is described. Numerical results compare favorably with experimental observations in terms of intersonic sliding tip speed, crack-like and pulse-like sliding modes and the stress fields at the sliding tip.

1 INTRODUCTION

Frictional sliding along an interface between two deformable solids is a basic problem of mechanics that arise in a variety of contexts including, for example, material processing, deformation and failure of fiber reinforced composites and earthquake dynamics. The classical Coulomb type of frictional relation relates the shear stress to the normal stress by a proportionaliy constant μ which can have a dependence on the relative sliding velocity. However, Adams (1995) and Ranjith and Rice (2001) showed that the problem of frictional sliding along an interface between two elastic solids, with sliding governed by Coulomb friction, is unstable to perturbations and hence ill-posed for a significant range of values of μ. Rate- and state-dependent models of friction have been introduced (Dieterich, 1979; Rice and Ruina, 1983; Ruina, 1983) that phenomenologically characterize the surface evolution and provide a representation of the transition from static to dynamic friction at constant normal load. For varying normal stress, Prakash and Clifton (1993) added an additional state variable to account for their observation of a delay in the change in shear traction following a sudden change in the normal traction. The use of these friction laws regularizes the sliding friction problem of elastic bodies with changing normal stress and make it a well-posed problem. We report on the implementation of a rate-state friction model in a finite-element code to simulate the frictional sliding behavior of two elastic solids under dynamic loading conditions. The computational results of dynamic sliding are compared with experimental observations.

2 FORMULATION AND NUMERICS

In a finite strain Lagrangian formulation, with the initial undeformed configuration taken as reference and with all field quantities considered to be functions of convected coordinates y^i and time t, the principle of virtual work can be written as

$$\int_V \mathbf{S} : \delta\mathbf{E}\,dV - \int_{S_{int}} \mathbf{T} \cdot \delta[\mathbf{u}]\,dS = \int_{S_{ext}} \mathbf{T} \cdot \delta\mathbf{u}\,dS - \int_V \rho \frac{\partial^2 \mathbf{u}}{\partial t^2} \cdot \delta\mathbf{u}\,dV \qquad (1)$$

where $\mathbf{S}$ is the second Piola-Kirchhoff stress tensor, $\mathbf{u}$ is the displacement vector $[\mathbf{u}]$, is the displacement jump across the cohesive surface, $\mathbf{A} : \mathbf{B}$ denotes $A^{ij}B_{ji}$, and V, S_{ext} and S_{int} are the volume, external surface area and internal cohesive or frictional surface area, respectively, of the body in the reference configuration. The density of the material in the reference configuration is ρ, $\mathbf{T}$ is the traction vector and the Lagrangian strain, $\mathbf{E}$, is given by

$$\mathbf{E} = \frac{1}{2}(\mathbf{F}^T \cdot \mathbf{F} - \mathbf{I}), \qquad \mathbf{F} = \mathbf{I} + \frac{\partial \mathbf{u}}{\partial \mathbf{x}} \qquad (2)$$

with $\mathbf{I}$ the identity tensor and $\mathbf{x}$ denoting the position vector of a material point in the reference configuration.

A configuration studied is sketched in Fig. 1. At $t = 0$, the body is at rest, an initial compressive stress is prescribed and impact loading is imposed. The calculations are two dimensional assuming plane stress conditions. A frictional constitutive relation is prescribed along the centerline and the material is taken to be linear elastic.

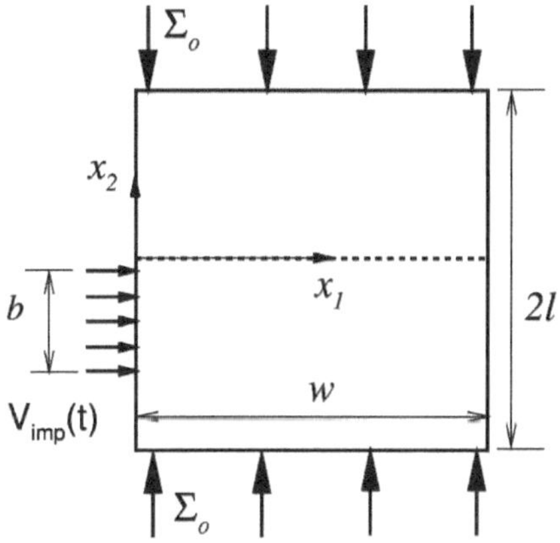

Figure 1: Geometry and loading configuration used in the finite element calculation.

The friction constitutive law used is a modified version of the Prakash-Clifton rate- and state-dependent law (Coker et al., 2004) characterized by the relation,

$$T_s = \mu(\theta_0, V_{slip})(\theta_1 + \theta_2) \tag{3}$$

where V_{slip} is the relative sliding velocity of the interface, T_s is the shear traction at the interface, θi are the state variables characterizing the history of the interface. The evolution equations for the state variables also include a characteristic length and the parameters are chosen so that the shear traction decreases with sliding velocity (velocity-weakening) as shown in Fig. 2a. The behavior in the apparent coefficient of friction, T_s/T_n that this law exhibits when subjected to a step jump in sliding velocity is shown in Fig. 2b.

A finite element discretization is used that is based on linear displacement triangular elements that are arranged in a "crossed-triangle" quadrilateral pattern. The equations of motion that result from substituting the finite element discretization into (1) are integrated numerically by an explicit integration procedure, the Newmark β-method with $\beta = 0$, Belytschko et al. (1976). A lumped mass matrix is used instead of the consistent mass matrix, since this has been found preferable for explicit time integration procedures, from the point of view of accuracy as well as computational efficiency, Krieg and Key (1973).

The experimental procedures used for the configuration sketched in Fig. 1 are similar to those used to study shear crack propagation (Rosakis et al., 1999). The impact loading is imposed via a cylindrical steel projectile fired using a gas gun with impact speeds ranging from 10 m/s to 60 m/s. Dynamic photoelasticity is used to extract stress field information around the interface. The photoelastic fringe patterns (isochromatic fringes that are the contours of the difference of maximum and minimum in-plane principal stresses) were recorded in real time

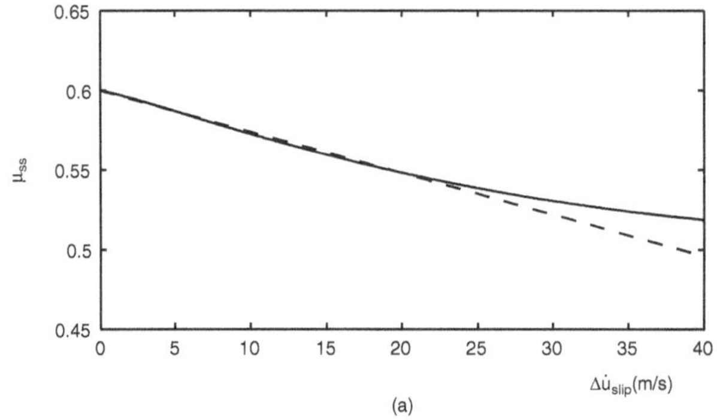

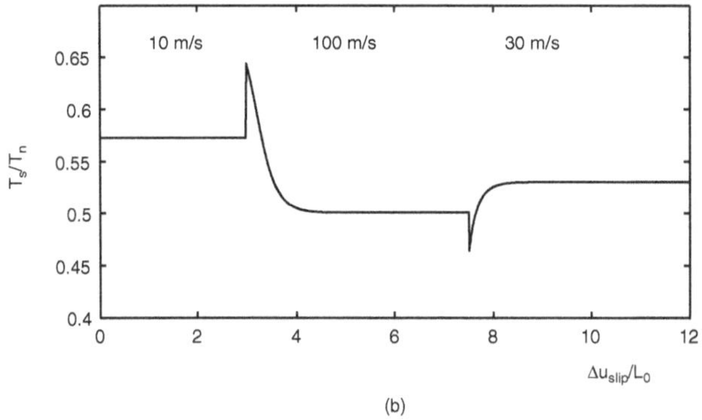

Figure 2: (a) Steady-state coefficient of friction, μ_{ss}, as a function of the sliding rate, $\Delta\dot{u}_{slip}$. (b) The effect of an abrupt change in the sliding rate on the apparent coefficient of friction for the rate- and state-dependent friction relation used in the calculations.

using a high-speed Cordin CCD camera capable of capturing 16 images at a rate of 100 million frames per second.

3 RESULTS

Finite element calculations using rate- and state-dependent friction laws for the interface showed that sliding of an interface can occur in different modes depending on the imposed compressive stress and the impact conditions. Friction modes that involve uniform sliding behind the rupture front, an isolated slip pulse, multiple slip pulses or a combination of these were obtained. It is found that the isolated slip pulse tends to become narrower and steeper with time while the multiple pulses do not. The experiments also exhibited both pulse-like and crack-like modes of sliding with a transition from crack-like mode to pulse-like mode with increasing impact speeds. In both experiments and calculations Mach lines in the photoelastic stress contours are observed. Calculations from the angle of the Mach lines show that the sliding tip travels at a speed greater than $\sqrt{2}\,c_s$ and that a trailing pulse travels faster than c_l. The results are presented in Coker et al. (2004).

ACKNOWLEDGEMENTS

DC and AN are pleased to acknowledge support from the Office of Naval support from the Research through grant N00014-97-1-0179 and from the General Motors Cooperative Research Laboratory at Brown University. GL and AJR are grateful for support from the Office of Naval Research through grant N00014-02-1-0522.

References

[1] Adams, G.G., 1995. Self-excited oscillations of two elastic half-spaces sliding with a constant coefficient of friction. J. Appl. Mech. 62, 867–872.

[2] Belytschko, T., Chiapetta, R.L. and Bartel, H.D., 1976. Efficient large scale non-linear transient analysis by finite elements. Int. J. Numer. Meth. Engr., 10, 579–596.

[3] Coker, D., Lykotrafitis, G, Needleman, A. and Rosakis, A.J., 2004. Frictional sliding modes along an interface between identical elastic plates subject to shear impact loading. To be published.

[4] Dieterich, J.H., 1979. Modeling of rock friction 1. Experimental results and constitutive equations. J. Geophys. Res. 84, 2161–2168.

[5] Krieg, R.O. and Key, S.W., 1973. Transient shell response by numerical time integration. Int. J. Numer. Meths. Engrg., 7, 273–286.

[6] Prakash, V., Clifton, R.J., 1993. Pressure-shear plate impact measurement of dynamic friction for high speed machining applications. Proc. 7th Int. Congress on Exp. Mech., Society of Experimental Mechanics, Bethel, CT, pp. 556–564.

[7] Ranjith, K., Rice, J.R., 2001. Slip dynamics at an interface between dissimilar materials, J. Mech. Phys. Solids, 49, 341–361.

[8] Rice, J.R., Ruina, A.L., 1983. Stability of frictional sliding. J. Appl. Mech., 50, 343–349.

[9] Rosakis, A.J., Samudrala, O., Coker, D., 1999. Cracks faster than the shear wave speed. Science, 284, 1337–1340.

[10] Ruina, A.L., 1983. Slip instability and state variable friction laws. J. Geophys. Res., 88, 10359–10370.

2 Structural Failure and Acoustic Emission

2.1 AE theoretical studies

Earthquakes and Acoustic Emission – Carpinteri & Lacidogna (eds)
© 2007 Taylor & Francis Group, London, ISBN 978-0-415-44402-6

IDENTIFICATION OF MICROCRACKING PROCESS IN FRACTURE PROCESS ZONE BY AE

M. Ohtsu
Graduate School of Science and Technology, Kumamoto University, Kumamoto 860-8555, Japan

ABSTRACT

Microcracking process in concrete is readily monitored by acoustic emission (AE) technique. Crack kinematics associated with nucleation of the fracture process zone are identified by AE-SiGMA procedure, by which crack locations, crack types and crack orientation are determined from the moment tensor analysis of AE waveforms.

By applying SiGMA, microcracking process in the process zone is investigated in a notched specimen of concrete. A relation between the damage variable in damage mechanics and the moment tensor is derived. The damage evolution in the fracture process zone is estimated under off-center bending. It is found that the damage evolve gradually in mode I failure along with mode-II failure and mixed-mode. Since the direction of crack extension can be derived from the eigenvectors of the moment tensor, the concept of the maximum circumferential stress is applied to estimate the normalized stress intensity factors. In the beginning of microcracking process, mode I failure is dominant. Then, mode II cracks are observed in the final stage. This result in SiGMA analysis is confirmed by BEM (boundary element method) analysis.

1 INTRODUCTION

The generalized theory of acoustic emission (AE) was established on the basis of elastodynamics (Ohtsu and Ono [1]). Although AE waves due to microcracking are discussed in concrete, theoretical treatment is associated with elastic waves in a homogeneous medium (Ohtsu [2]). This is because elastodynamic properties of material constituents are physically dependent on the relation between the wavelengths and the characteristic dimensions of heterogeneity. In the case that the wavelengths are even larger than the sizes of heterogeneous inclusions, the effect of heterogeneity is inconsequent. This is the case of concrete, if the sizes of specimens are large enough compared with the wavelengths.

Crack kinematics of AE source are defined by a crack motion vector (Burgers vector) and a normal vector to a crack plane. These two vectors could leads to the moment tensor analysis (Ohtsu et al. [3]). From the moment tensor, quantitative information on crack kinematics of crack locations, crack types and crack orientations is obtained. In order to determine moment tensor components from AE waveforms, SiGMA procedure (simplified Green's functions for moment tensor analysis) is developed (Ohtsu [4]). AE sources are classified into tensile cracks and shear cracks by applying the eigenvalue analysis to the moment tensor, and the direction of crack motion is derived from the eigenvectors.

In this paper, a relation between the damage variable in damage mechanics and the moment tensor is derived. Then, the damage evolution in the fracture process zone of a notched concrete beam is estimated under off-center bending. Since the direction of crack extension can be derived from the eigenvectors, the normalized stress intensity factors are estimated experimentally.

2 MOMENT TENSOR AND DAMAGE EVOLUTION

An elastodynamic solution of wave motion $\mathbf{u}(\mathbf{x},t)$ due to cracking is mathematically represented,

$$\begin{aligned} u_k(\mathbf{x},t) &= \int_F G_{kp,q}(\mathbf{x},\mathbf{y},t)\, C_{pqij}\, n_j * b_i(\mathbf{y},t)dF \\ &= G_{kp,q}(\mathbf{x},\mathbf{y},t) * S(t)\, C_{pqij}\, n_j l_i \int_F b(y)dF \\ &= G_{kp,q}(\mathbf{x},\mathbf{y},t) * S(t)\, C_{pqij}\, n_j l_i \Delta V, \end{aligned} \tag{1}$$

where $b(\mathbf{y},t)$ is the crack motion vector (Burgers vector) and the asterisk symbol * represents the convolution integral in time. C_{pqji} are the elastic constants, and $G_{ip,q}(\mathbf{x},\mathbf{y},t)$ are the spatial derivatives of Green's functions as they imply $\partial G_{ip}(\mathbf{x},\mathbf{y},t)/\partial x_q$. $\mathbf{n}$ is the normal vector to the crack surface, and $\mathbf{l}$ is the unit direction vector of the crack motion $b(\mathbf{y})$. $S(t)$ is the source-time function of crack kinetics and ΔV is the crack volume.

Since eqn (1) is fairly complicated and contains two vectors $\mathbf{l}$ and $\mathbf{n}$, it is not suitable for an inverse problem. Thus, introducing moment tensor M_{pq}, Eqn (1) is simplified as eqn (3),

$$\int_F C_{pqkl}\, b(\mathbf{y}) l_k\, n_l\, dF = C_{pqkl} l_k n_l \int_F b(\mathbf{y})dF = C_{pqkl} l_k n_l\, \Delta V = M_{pq} \tag{2}$$

$$u_k(\mathbf{x},t) = G_{kp,q}(\mathbf{x},\mathbf{y},t)\, M_{pq} * S(t). \tag{3}$$

The moment tensor, M_{pq}, is defined by the product of the elastic constants [N/m^2] and the crack volume [m^3], which leads to the moment as physical unit [Nm]. In the case of an isotropic material,

$$M_{pq} = \lambda l_k n_k \delta_{pq} + 2\mu(l_p n_q + l_q n_p)\Delta V, \tag{4}$$

where λ and μ are Lame constants.

In damage mechanics (Kachanov [5]), damage tensor D_{pq} is defined as (Ohtsu and Ohtsuka [6]),

$$D_{pq} = (l_p n_q + l_q n_p)\Delta V/(2V^*). \quad (5)$$

Here, V* is the representative volume. Then, the scalar damage variable is derived,

$$D = n_p D_{pq} n_q = l_k n_k \Delta V/V^*. \quad (6)$$

From eqn (4), a trace component is obtained,

$$M_{kk} = (3\lambda + 2\mu) l_k n_k \Delta V. \quad (7)$$

Comparing Eqn (7) with Eqn (6), it is found that the trace component of the moment tensor is equivalent to the scalar damage variable. Consequently, damage evolution can be estimated from the summation of the trace components of the moment tensors. In addition, the accumulation of crack volumes is also relatively estimated from $M_{kk}/l_k n_k$.

3 SIGMA PROCEDURE

For the analysis of the inverse problem for eqn (3), the spatial derivatives of Green's functions are inevitably required. Accordingly, numerical solutions are obtained by FDM (Enoki et al. [7]) and by FEM (Hamstad et al. [8]). These solutions, however, need a vector processor for computation and are not readily applicable to processing a large amount of AE waves. Based on the far-filed term of P wave, a simplified procedure was developed, which is suitable for a PC-based processor and robust in computation. The procedure is now implemented as SiGMA (Simplified Green's functions for Moment tensor Analysis) code (Ohtsu [4]).

By taking into account only the far field term of Green's functions in an infinite space, the displacement $U_i(\mathbf{x},t)$ of P wave motion is obtained from Eqn (1),

$$U_i(\mathbf{x},t) = -1/(4\pi\rho v_p^{\,3})(r_i r_p r_q/R)\, dS(t)/dt\, M_{pq}. \quad (8)$$

Here ρ is the density of the material and v_p is P-wave velocity. R is the distance between the source y and the observation point x, of which direction cosine is $\mathbf{r} = (r_1, r_2, r_3)$. Considering the effect of reflection at the surface and neglecting the source-time function, amplitude $A(\mathbf{x})$ of the first motion is represented,

$$A(x) = Cs\, Ref(\mathbf{t},\mathbf{r})/R\; r_p M_{pq} r_q \quad (9)$$

where Cs is the calibration coefficient including material constants in Eqn (8). **t** is the direction of the sensor sensitivity. Ref(**t**,**r**) is the reflection coefficient at the observation location **x**. Since the moment tensor is symmetric, the number of independent unknowns M_{pq} to be solved is six. Thus, multi-channel observation of the first motions at more than six channels is required to determine the moment tensor components.

From AE waveform in Figure 1, two parameters of the arrival time (P1) and the amplitude of the first motion (P2) are determined. In the source location procedure, source location y is determined from these arrival time differences. Then, distance R and its direction vector **r** are determined. The amplitudes of the first motions at more than 6 channels are substituted into Eqn 9, and the components of the moment tensor are determined. Since SiGMA code requires only relative values of the moment tensor components, the relative calibration of AE sensors is sufficient enough. Then, the classification of a crack is performed by the eigenvalue analysis of the moment tensor. Setting the ratio of the maximum shear contribution as X, three eigenvalues for the shear crack become X, 0, −X. Likewise, the ratio of the maximum deviatoric tensile component is set as Y and the isotropic tensile as Z. It is reasonably assumed that the principal axes of the shear crack is identical to those of the tensile crack. Then, the eigenvalues of the moment tensor for a general case are represented by the combination of the shear crack and the tensile crack. Thus, the eigenvalues are normalized and decomposed,

$$\begin{aligned} 1.0 &= X + Y + Z, \\ \text{the intermediate eigenvalue/the maximum eigenvalue} &= 0 - Y/2 + Z, \quad (10) \\ \text{the minimum eigenvalue/the maximum eigenvalue} &= -X - Y/2 + Z. \end{aligned}$$

Here X, Y, and Z denote the shear ratio, the deviatoric tensile ratio, and the isotropic tensile ratio, respectively. In SiGMA code, AE sources of which the shear ratios are less than 40% are classified into tensile cracks. The sources of $X > 60\%$ are classified into shear cracks. In between 40% and 60%, cracks are referred to as mixed mode.

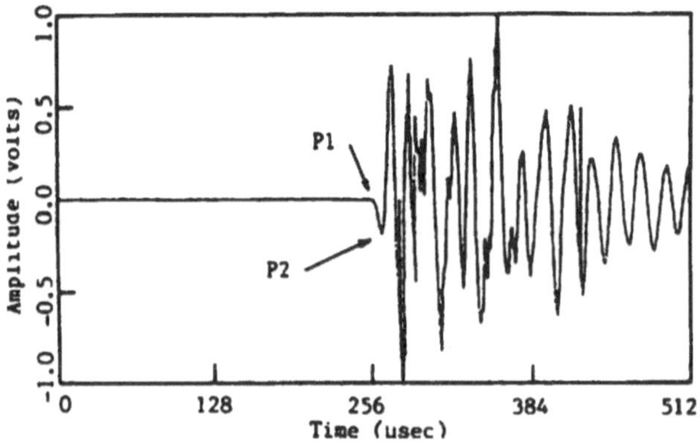

Figure 1: AE waveform.

4 MOMENT TENSOR AND MIXED-MODE CRACKING

Erdgan and Sih [9] proposed the maximum circumferential stress criterion for the mixed-mode crack extension. As given in Figure 2, the direction of crack extension to a pre-existing crack is determined as the direction of the maximum circumferential stress,

$$K_I \sin\theta + K_{II}(3\cos\theta - 1) = 0. \tag{11}$$

Where K_I and K_{II} are the stress intensity factors of mode I and mode II, respectively. Initiation of the crack is governed by,

$$\cos\theta/2[K_I \cos^2\theta/2 - 3/2\, K_{II} \sin\theta] = K_{IC}. \tag{12}$$

Here K_{IC} is the critical stress intensity factor. From eqns (11) and (12), normalized stress intensity factors $K_I{}^* = K_I/K_{IC}$ and $K_{II}{}^* = K_{II}/K_{IC}$ are derived as,

$$\begin{aligned} K_I^* &= (3\cos\theta - 1)/[\cos\theta/2(\cos\theta + 1)], \text{ and} \\ K_{II}^* &= -\sin\theta/[\cos\theta/2(\cos\theta + 1)] \end{aligned} \tag{13}$$

In the eigenvalue analysis, three eigenvectors are also determined, and then the vectors **l** and **n**, which are interchangeable, are recovered. Three eigenvectors **e1**, **e2**, and **e3** are obtained as,

$$\begin{aligned} \mathbf{el} &= \mathbf{l} + \mathbf{n} \\ \mathbf{el} &= \mathbf{l} \times \mathbf{n} \\ \mathbf{el} &= \mathbf{l} - \mathbf{n}. \end{aligned} \tag{14}$$

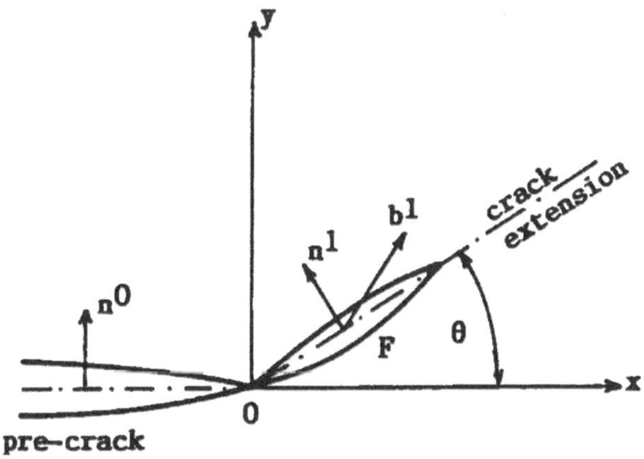

Figure 2: Crack extension from a pre-crack.

Here x denotes the vector product. From these relations, the two vectors **l** and **n** associated with the orientation of a crack can be recovered. According to Figure 2, $\cos\theta = \mathbf{n}^0 \cdot n^1$, and $\mathbf{n}^1$ is identical to vector **n** in Eqn (14). Consequently, normalized stress intensity factors in eqn (13) can be determined experimentally from SiGMA analysis (Ohtsu et al. [10]).

5 MICROCRACKING PROCESS IN FRACTURE PROCESS ZONE

In a three-point bending test of an off-centered-notched concrete beam, SiGMA analysis was performed. Results are given in Figure 3. The tensile crack is indicated by arrow symbol, while the shear crack is denoted by cross symbol. The directions of two vectors **l** and **n** are shown in their directions. It is found that both types of tensile cracks and shear cracks are observed in the fracture process zone ahead of the notch. Then, damage evolution was estimated by Eqn (7). It is realized that the evolution process is fairly gradual, nucleating all types of tensile, shear and mixed-mode cracks.

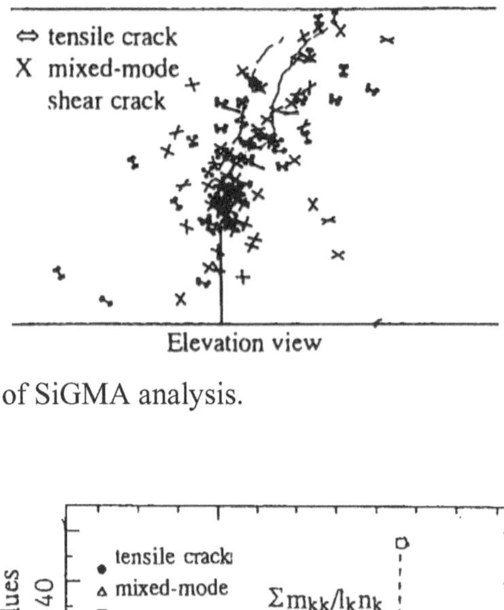

Figure 3: Results of SiGMA analysis.

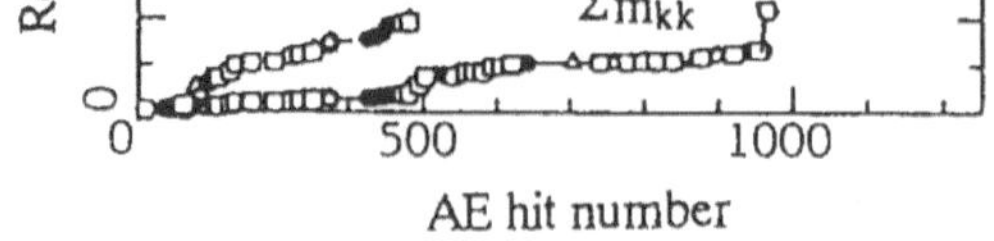

Figure 4: Evolution of damage in the process zone.

To investigate the microcracking process, analysis of the two-domain boundary element method (BEM) was performed. Crack traces observed in the test and the analysis are compared in Figure 5. It is found that the crack trace is reasonably simulated by BEM.

Normalized stress intensity factors (SIF) are then calculated from results of SiGMA analysis and are shown in Figure 6. These are plotted against AE hit number. In the beginning stage of microcracking process, K_I* is dominant. Then, K_{II}*-dominant cracks are observed in the final stage. In order to confirm this

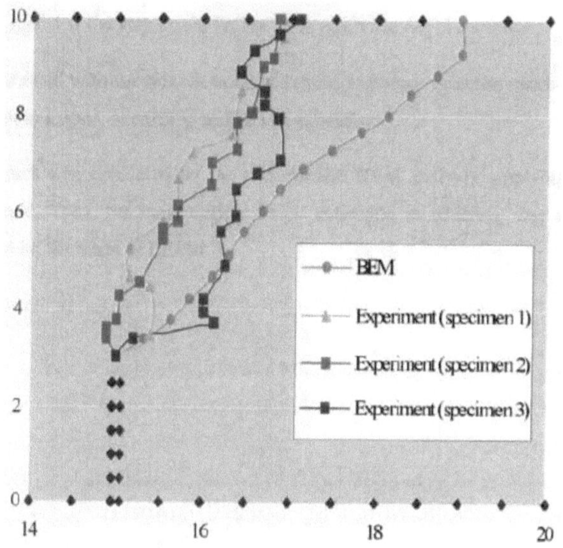

Figure 5: Crack traces observed in the test and analyzed by BEM.

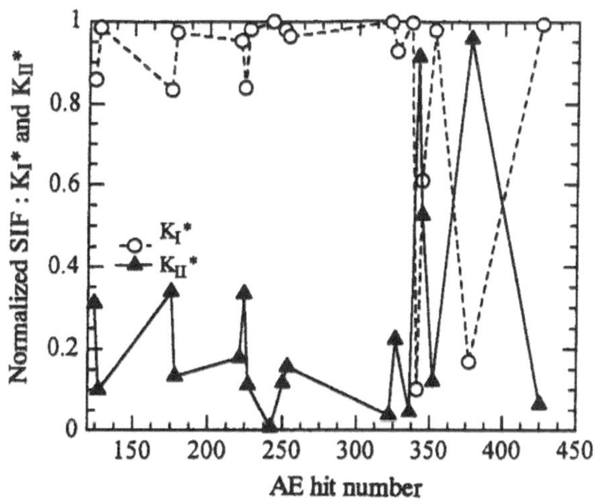

Figure 6: Normalized SIF by SiGMA analysis.

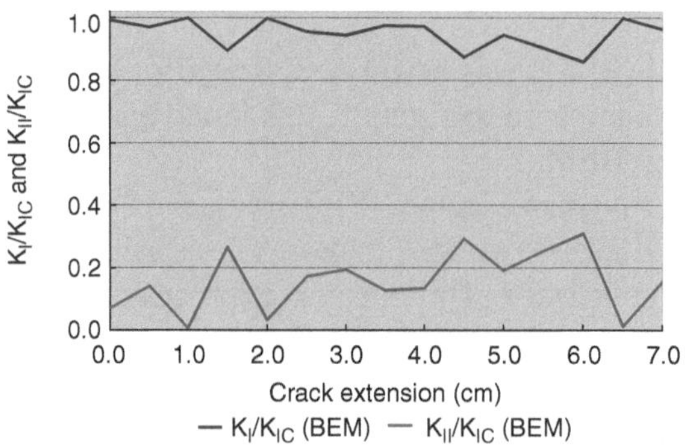

Figure 7: Normalized SIF by BEM analysis.

result, normalized SIFs were determined from BEM analysis. Results are shown in Figure 7. Under crack extension, K_I* decreases and K_{II}* increases. Thus, the tendency observed in SiGMA analysis is demonstrated. It is concluded that the moment tensor analysis is very promising to quantitatively identify microcracking processes in the fracture process zone of concrete.

References

[1] Ohtsu, M. and Ono, K. "A Generalized Theory of Acoustic Emission and Green's Functions in a Half Space", Journal of AE, Vol. 3, No. 1, 124–133, 1984.

[2] Ohtsu, M. "Source Mechanism and Waveform Analysis of Acoustic Emission in Concrete", Journal of AE, Vol. 2, No. 1, 103–112, 1982.

[3] Ohtsu, M., Okamoto, T. and Yuyama, S. "Moment Tensor Analysis of Acoustic Emission for Cracking Mechanisms in Concrete", ACI Structural Journal, Vol. 95, No. 2, 87–95, 1998.

[4] Ohtsu, M. "Acoustic Emission Theory for Moment Tensor Analysis", Research in Nondestructive Evaluation, No. 6, 169–184, 1995.

[5] Kachanov, M. "Effective Elastic Properties of Cracked Solids", Applied Mechanics Review, Vol. 45, No. 8, 304–335, 1992.

[6] Ohtsu, M. and Ohtsuka, M. "Damage Evolution by AE in the Fracture Process Zone of Concrete", J. Materials, Conc. Struct. Pavement, JSCE, No. 599/V-40, 177–184, 1998.

[7] Enoki, M., Kishi, T. and Kohara, S. "Determination of Micro-cracking Moment Tensor of Quasi-cleavage Facet by AE Source Characterization", Progress in Acoustic Emission III, JSNDI, 763–770, 1986.

[8] Hamstad, M.A., O'Gallagher, A. and Gary, J. "Modeling of Buried Monopole and Dipole Source of Acoustic Emission with a Finite Element Technique", Journal of AE, Vol. 17, Nos. 3–4, 97–110, 1999.

[9] Erdogan, F. and Sih, G.C. "On the Crack Extension in Plates under Plane Loading and Transverse Shear", J. Basic Engineering, Transactions of ASME, No. 12, 519–527, 1963.

[10] Ohtsu, M., Munwam, M.C. and Rossmanith, H.P. "Crack Extension in Cementitious Materials by AE", Concrete Science and Engineering, Vol. 1, 157–165, 1999.

Earthquakes and Acoustic Emission – Carpinteri & Lacidogna (eds)
© 2007 Taylor & Francis Group, London, ISBN 978-0-415-44402-6

CONSEQUENCES OF ACOUSTIC EMISSION ON CRACK SPEED AND ROUGHNESS EXPONENT IN BRITTLE DYNAMIC FRACTURE

A. Parisi[1] & R.C. Ball[2]
[1]Laboratoire PMC,' Ecole Polytechnique, 91128 Palaiseau, France
[2]Department of Physics, University of Warwick, Coventry CV4 7AL, UK

ABSTRACT

We show by computer simulations that acoustic emission from the crack tip strongly reduces the delivery of fracture work, due to the coupling between the crack speed and the acoustic branches in dispersive media. The direct consequence is a selection criterion for the terminal crack speed which, for planar cracks, produces results corresponding to those found in experiments on highly anisotropic materials. In case of isotropic material with cracks of unrestricted geometry, the drop in the crack speed with respect to the planar case is connected to a mechanism of attempted branching, which is also responsible for the logarithmic roughness of the final fracture for marginal loadings. Higher loadings lead to a well defined roughness exponent of $\zeta \sim 0.45$ compatible with that measured experimentally at short length scales, and in our simulations clearly connected with the generation of macroscopic branches.

1 INTRODUCTION

Acoustic emission is commonly used to analyze the precursors of fracture, exploiting the possibility to determine the location, frequency and energy of micro-fracturing events. Less common is the analysis of acoustic emission from moving cracks because locating a crack and measuring the energy involved is better achieved by different methods. An interesting aspect of the acoustic emission in dynamic fracture is that this phenomenon is only partially covered by continuum theory. Continuous media support the propagation of waves travelling at the characteristic sound speeds v_l and v_t of the longitudinal and transversal acoustic modes, and at the Rayleigh speed v_R for surface waves. The prediction of the continuum theory for the terminal crack speed is that a crack never exceeds v_R [1]. Up to that speed, it describes an advancing crack as a continuous deformation of the medium and no sound emission is expected because there can be no coupling between the crack speed and sound waves. Experiments in both real materials and simulations nevertheless show intense acoustic emission [2, 3], which cannot be directly accounted for in the continuum limit. This emission can be understood as a consequence of discreteness which is always present in real materials (and in simulations) and is masked in the theory by the continuum approach [4]. In the following we show that such acoustic emission influences the amount of energy which goes into fracture work depending on the crack speed. This leads to a speed selection criterion for crack propagation and influences the scaling of the final surface.

2 DISCRETENESS AND ENERGY RELEASE RATE

In presence of discreteness and hence of acoustic emission, the macroscopic energy release rate provided by the continuum theory corresponds to the sum of two distinct contributions [4]:

$$G_M(v,t) = G_{\text{br}}(v,t) + G_{\text{ph}}(v,t) \tag{1}$$

Here $G_M(v,t)$ is the solution of the continuum limit which governs the delivery of energy towards the crack tip and hence represents the energy *available* at the crack tip at a given time t. $G_{\text{br}}(v,t)$ is the actual *breakage* energy release rate, or the portion of the available energy which effectively goes locally into fracture work, whilst $G_{\text{ph}}(v,t)$, the *phonon* energy release rate, is the portion of the available energy which is radiated as acoustic emission. $G_M(v,t)$ can be obtained from the continuum theory for a given (computer) experimental setup, whilst $G_{\text{br}}(v,t)$ can be measured in simulations by fixing the crack speed and measuring the resulting fracture work.

We have introduced a novel finite element model which permits fast three-dimensional simulations and is amenable of explicit analytical treatement [4]. The scheme adopted introduces a discretization of the continuum elastodynamic equations by using an fcc lattice of massive sites. The lattice geometry is then reflected in the acoustic properties of the material in that the materials is dispersive, and influences the behaviour of the energy release rate. We have introduced a novel finite element model which permits fast three-dimensional simulations and is amenable of explicit analytical treatement [4]. The scheme adopted introduces a discretization of the continuum elastodynamic equations by using an fcc lattice of massive sites. The lattice geometry is then reflected in the acoustic properties of the material in that the materials is dispersive, and influences the behaviour of the energy release rate.

By a series of computer simulations we measured the breakage energy release rate $G_{\text{br}}(v,t)$ for a planar crack, in the special case of a strip geometry with fixed displacements at the top and bottom boundaries. In this case the total available

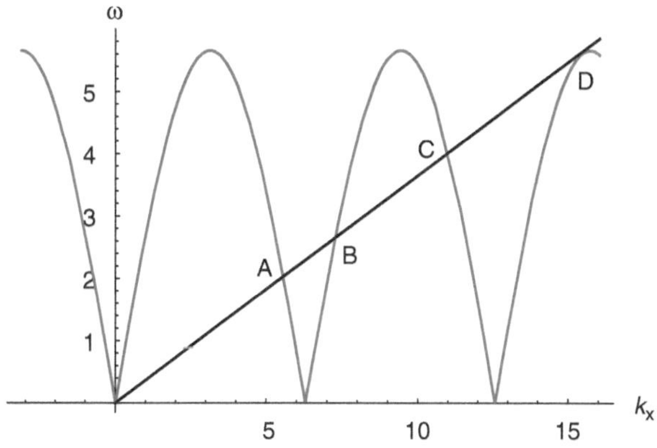

Figure 1: Schematic representation of the basic phenomenon. A periodic lattice generates a periodic acoustic band structure. A crack advancing at some speed v matches such periodic band structure at points A, B, C, and D in the example shown, emitting waves at the corresponding frequencies ω and wavevectors k_x. Resonant emission is found in D: in this case the crack speed v also matches the group velocity $\partial\omega/\partial k_x$ of emitted waves.

macroscopic fracture energy $G_M(v,t)$ is independent of the crack speed [1] and, for cracks longer than the height of the sample, all energy release rates of Eq. (1) are time independent, so that we get $G_M^\infty = G_{\rm br}(v) + G_{\rm ph}(v)$. Although the macroscopic energy release rate is independent of the crack speed, our simulations show that the breakage energy release rate does depend on the crack speed. Such speed dependence is a consequence of the coupling between the crack speed and the acoustic emission via the acoustic dispersion relations. Fig. 1 describes the basic phenomenon. At any crack speed $v < v_R$ the moving tip emits sound waves at the frequency and wavevector of the corresponding permitted acoustic modes. In particular, for some special speeds we have resonant emission when the emitted waves travel at the same speed as the crack itself, leading to an increase of the energy radiated into phonons. This in turn leads to a decrease in the energy going into fracture work and thus explains the speed dependence of the breakage energy release rate.

The relationship between the microscopic breakage energy release rate and the macroscopic energy release rate is better expressed by introducing the *efficiency* $E(v)$:

$$G_{\rm br}\left(\nu,t\right) = E\left(\nu\right) G_M\left(\nu,t\right) \tag{2}$$

In the fixed grip setup discussed above, this gives $G_{\rm br}(v) = E(v)\, G_M^\infty(v)$, so that the efficiency becomes the sole source of velocity dependence. We have shown [4] that $E(v)$ only depends on the lattice geometry and crack speed, being local

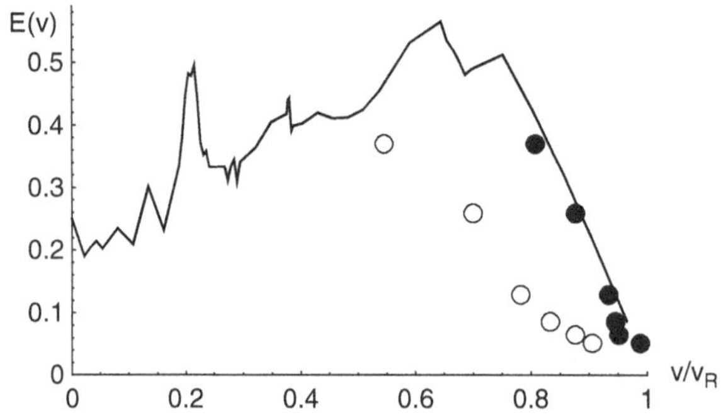

Figure 2: The continuous line shows the speed dependence of efficiency. Filled circles correspond to simulations of planar crack in a disordered medium for different loadings. The topmost point corresponds to the lowest loading and vice-versa: increasing the loading leads to an increase in the terminal crack speed. Empty circles correspond to simulations of non-planar cracks in a disordered medium; the reduced crack speed is associated with crack branching.

to the crack tip and thus independent on the macroscopic dynamical regime which instead is generally described by $G_M(v,t)$.

3 CRACK SPEED AND ROUGHNESS EXPONENT

The translation of the standard Griffith criterion in the framework of Eq. (1) is that a crack will advance only when $G_{\mathrm{br}}(v,t)$ exceeds a threshold value connected with the toughness of the material. In the fixed grip setup discussed above, this translates into a threshold for the efficiency $E(v)$. For a given threshold value for $E(v)$ the possible crack speeds are directly obtained by looking at its speed dependence in Fig. 2, and simple stability argument [4] leads to the conclusion that stable crack propagation is only possible when $E(v)$ is a decreasing function of the crack speed.

We have performed numerical simulations of moving cracks by fixing the threshold and varying the applied loading. In Fig. 2 full circles correspond to simulations of planar cracks in a disordered medium for different loadings: all points fall on the efficiency curve. Furthermore, for low loadings (corresponding in the figure to high values of $E(v)$) the crack speed is fully compatible with the value measured in highly anisotropic materials [5, 6]. Empty circles show the results on the terminal crack speed for nonplanar cracks, when out of plane breakage is allowed. We have verified [7] that the drop in crack speed, with respect to the planar case, is directly connected with a mechanism of *attempted branching*, when new branches try to open. In this situation, not all the energy is used to have the crack advance, and the crack slows down. On the other hand, the energy available is insufficient to initiate macroscopic branches,

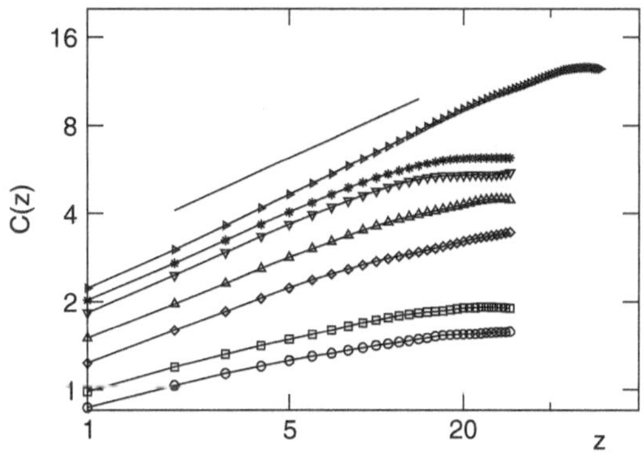

Figure 3: The height-height correlation function of the final fracture surface for increasing loadings. The highest curves have a slope similar to the reference slope $\zeta = 0.45$ shown by the continuous line. The lowest curves correspond to the lowest loadings: note that they deviate from the reference slope. It can be shown that for such loadings the scaling of the surface is logarithmic [4].

and the resulting topology is compatible with a logarithmic scaling, found in experiments of marginal loading [8] and in theoretical calculations [9, 10]. Only for higher loadings does the roughness exponent ζ grow to values of $\zeta \approx 0.45$ (see fig. 3) found in many experiments [11, 12, 13], and this is connected with the appearence of macroscopic branches in our simulations.

4 CONCLUSIONS

We have shown that acoustic emission for a moving cracks has a measurable effect on the dynamics, introducing a selection criterion for the crack speed which, in the case of a strip geometry with fixed displacement at the boundaries, is missing in the continuum theory. The introduction of the efficiency $E(v)$ provides a way to predict the terminal crack speed in any loading configuration, at least given a preferred direction for crack propagation, and gives results comparable to experiments in the setup investigated. The departure from this prediction when a preferred direction is missing, is connected to microbranching which is responsible for the logarithmic scaling at marginal loadings. The well defined roughness exponent $\zeta \approx 0.45$ is connected in our simulations to the growth to macroscopic sizes of such branches and is possible only for higher loadings.

References

[1] Freund, L. B.: Dynamic Fracture Mechanics. Cambridge University Press, 1990.

[2] Gross, S .P., Fineberg, J., Marder, M., McCormick, W. D., Swinney, H. L.: Phys. Rev. Lett., **71** 3162 (1993).
[3] Fratini, S., Pla, O., Gonzalez, P., Guinea, F., Louis, E.: Phys. Rev. B, **66** (10) 104104 (2002).
[4] Parisi, A., Ball, R. C.: Phys. Rev. B, **66**(16) 165432 (2002).
[5] Washabaugh, P. D., Knauss, W. G.: Int. J. Fracture, **65** 97 (1994).
[6] Fineberg, J., Marder, M.: Rev. Mod. Phys., **313** 1 (1999).
[7] Parisi, A., Ball, R. C.: Submitted to Phys. Rev. B, preprint: cond-mat/0403638.
[8] Ball, R. C., Larralde, H.: Int. J. Fracture, **71**(4) 365 (1995).
[9] Larralde, H., Ball, R. C.: Europhys. Lett., **30**(2) 87 (1995).
[10] Ramanathan, S., Ertas, D.: Fisher, D. S., Phys. Rev. Lett., **79** 873 (1997).
[11] Milman, V. Y., Blumenfeld, R., Stelmashenko, N. A., Ball, R. C.: Phys. Rev. Lett., **71** 204 (1993).
[12] Daguier, P., Nghiem, B., Bouchaud, E., Creuzet, F.: Phys. Rev. Lett., **78**(6) 1062 (1997).
[13] Bouchaud, E.: J. Phys.: Cond. Matt., **9** 4319 (1997).

Earthquakes and Acoustic Emission – Carpinteri & Lacidogna (eds)
© 2007 Taylor & Francis Group, London, ISBN 978-0-415-44402-6

FAILURE TIME, CRITICAL BEHAVIOUR AND ACTIVATION PROCESSES IN CRACK FORMATION

S. Ciliberto[1], S. Deschanel[2], A. Guarino[3], S. Santucci[1], R. Scorretti[4] & L. Vanel[1]
[1]Laboratoire de Physique, CNRS UMR5672, Ecole Normale Supérieure de Lyon, 46 allée d'Italie, 69364 Lyon, France
[2]GEMPPM, INSA de Lyon, 69621 Villeurbanne, France
[3]Université de Polynésie Française – BP 6570 FAA'A, Tahiti, French Polynesia
[4]Ecole Centrale de Lyon, 69134 Ecully, France

In this seminar we will review several experimental and theoretical aspects of two very important problems of material failure: crack precursors and crack prediction. We will summarize the results of several experiments and numerical simulations that we have performed in order to give new insight on these two important problems. Specifically the acoustic emission of fracture pre-cursors, and the failure time of samples made of heterogeneous materials (wood, fiber-glass) are studied as a function of the load features and geometry [1, 2, 3, 4]. The accurate study of the localization and of the statistics of acoustic emission is indeed very important in order to compare the experimental results with those of percolation and critical models for crack formation. A typical example of microcrack localization in a large cylindrical sample is described in Fig.1 Once the microcracks have been localized we study the statistics of the time interval δt between events (precursors) and the acoustic energy ε of each event. We find that they are power law distributed and that the exponents of these power laws depend on the load history and on the material. An example is shown in Fig.2 [1, 2, 3, 4].

When the sample failure is produced by an imposed stress (see Fig.3a) the cumulative acoustic energy E, that is the integral of ε as a function of time, presents a critical divergence near the failure time τ, which is

$$E \sim \left(\frac{\tau - t}{\tau} \right)^{-\gamma} \tag{1}$$

It is important to point out that this critical behavior is observed only if the stress is imposed, whereas no power law is observed when the strain is imposed, see Fig.3b). Notably the positive exponent γ, in Eq.1, is independent, within error bars, on all the experimental parameters (loading speed and materials). The failure time is, of course a function of the applied stress, which can be constant or a

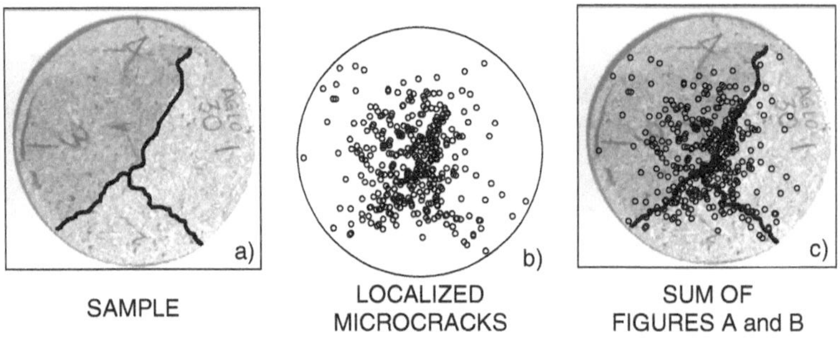

Figure 1: Microcrack localization Example of localization in a cylindrical sample of diameter 20 cm and thickness 0.3 cm. The samples are broken by a stress perpendicular to the base of the cylinder [1, 2, 3, 4]. In a) the broken wood sample is shown. We clearly distinguish the main final cracks. When a sample is submitted to an external stress, sound is emitted by microcracks appearing before the final failure. In b) the small circles represent the sources of acoustic waves detected and localized by mean of four detectors. In c), we superpose the figure a) and b). We clearly see that the density of acoustic sources (microcracks) is very large on the final cracks. The final crack is produced by the nucleation of many microcracks. This picture is qualitatively similar to that of percolation models for cracks.

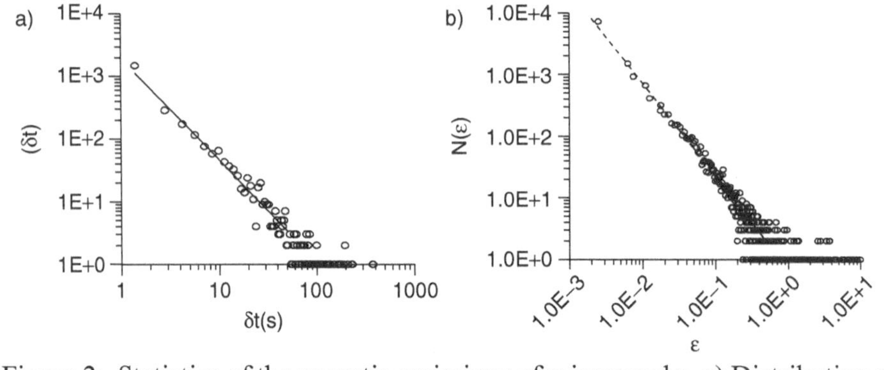

Figure 2: Statistics of the acoustic emissions of microcracks. a) Distribution of the time δt between two acoustic emissions occurring during the slow crack formation. b) distribution of the acoustic energy ε emitted by each microcrack.

function of time. If a constant pressure P is applied to a sample, we found that in heterogeneous materials the failure time τ has a functional dependence on P which is predicted with a good accuracy by a model of microcrack nucleation proposed by Pomeau [3, 4], specifically:

$$\tau = \tau_0 \exp\left(\frac{P_0}{P}\right)^4 \tag{2}$$

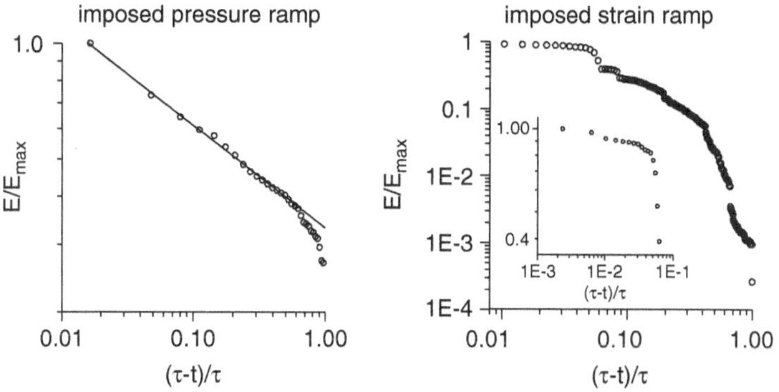

Figure 3: Cumulative acoustic energy of microcracks. The cumulative acoustic energy is plotted as a function of the reduced time $(\tau - t)/\tau$, where τ is the failure time of the sample. a) The sample is submitted to a linearly increasing pressure ramp. b) The sample is submitted to a linearly increasing strain ramp. We clearly see that in a) $E \sim (\tau - t/\tau)^{-\gamma}$. The exponent is independent within error bars by the loading speed and by the sample material. No critical behaviour is observed in b) when the strain is imposed.

where P is the external constant pressure applied to the sample τ and $P_0^4 \cong \Gamma^3 Y^2/(K_B T)$ is a parameter which depends on the temperature T, the Young modulus Y and the surface energy Γ of the material. A typical result of the experiments is shown in Fig.4, where the life time of a large cylindrical sample is plotted as a function of P^{-4}. We see the excellent agreement between the experimental points and the fit performed using Eq.2. We have extended this model to time dependent P (see ref.[3, 4] for details) and we find an excellent agreement with the experimental data. It is interesting to notice that even in the case when a constant pressure is applied to the sample the cumulative acoustic energy E has the dependence on time described by Eq.1. Thus the dependence of E on the reduced time $(\tau - t)/\tau$, shown in Fig.3, turns out to be a new type of critical behavior where the control parameter is only the time [4]. All the other parameter of the experiments are indeed taken into account by τ. In spite of the good agreement between the experimental data of τ and the functional dependence on P, given by Eq.2, the activation model fails to reproduce in a quantitative way the experimental results, mainly in the temperature dependence [3, 4].

Motivated by this discrepancy we have investigated in simple models (the fiber bundle model) the role of disorder on the fracture produced by a thermal activation of microcracks [5, 6]. We find that the presence of disorder (heterogeneity) contributes to increasing the effective temperature of a sample subject to a constant load. Equivalently, but perhaps more physically, one can state that disorder renormalizes the barrier height to be overcome in order to give rise to a macroscopic failure of the fiber bundle. This scenario can be understood by noticing that the fracture evolves through a sequence of many irreversible processes. After

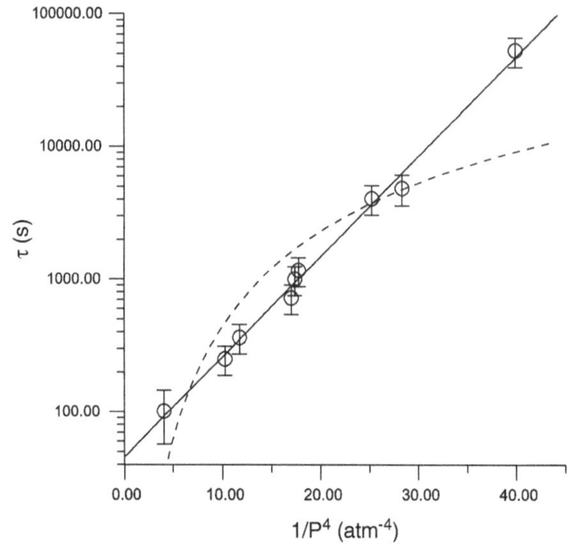

Figure 4: Life time of a sample submitted at a constant external pressure P. The life time τ is plotted as a function of P^{-4}. The continuous line is the fit with Eq.2. The dashed line is a best fit of the data using Mori or Zurkov laws. We clearly see that the fit with Eq.2 is much better than the other ones.

the failure of the weakest fibers, the system cannot any longer come back to its initial state, while, at the same time, the energy barrier has lowered. Although the fiber models explain the existence of a large effective temperature, they present a dependence on P which is quite different from that of Eq.2, which is not yet completely understood.

In order to give new insight into this problem we have studied, the role of an activation processes in the propagation of a single crack in a thin sheet of papers submitted to a tensile force F. The statistical and mean dynamical properties turn out to be in excellent agreement with the numerical and anaytical results of a 2D elastic model where crack growth is controlled by the thermal noise [7, 8]. In the experiment, we observe that the crack progresses by jumps as it can be seen in Fig.5a), where time is plotted as a function of the crack length L. The mean behaviour, obtained as the average of many experiments performed in the same conditions, is plotted in the insert of Fig.5a). The dotted line is the prediction of the 2D elastic model [7], that is

$$t = \tau \left[1 - \exp\left(\frac{L_i - L}{l_0} \right) \right] \tag{3}$$

where L_i is the initial crack length, l_0 a characteristic length, and τ is the life time of the sample. The experimental data plotted in the insert of Fig.5a) are in

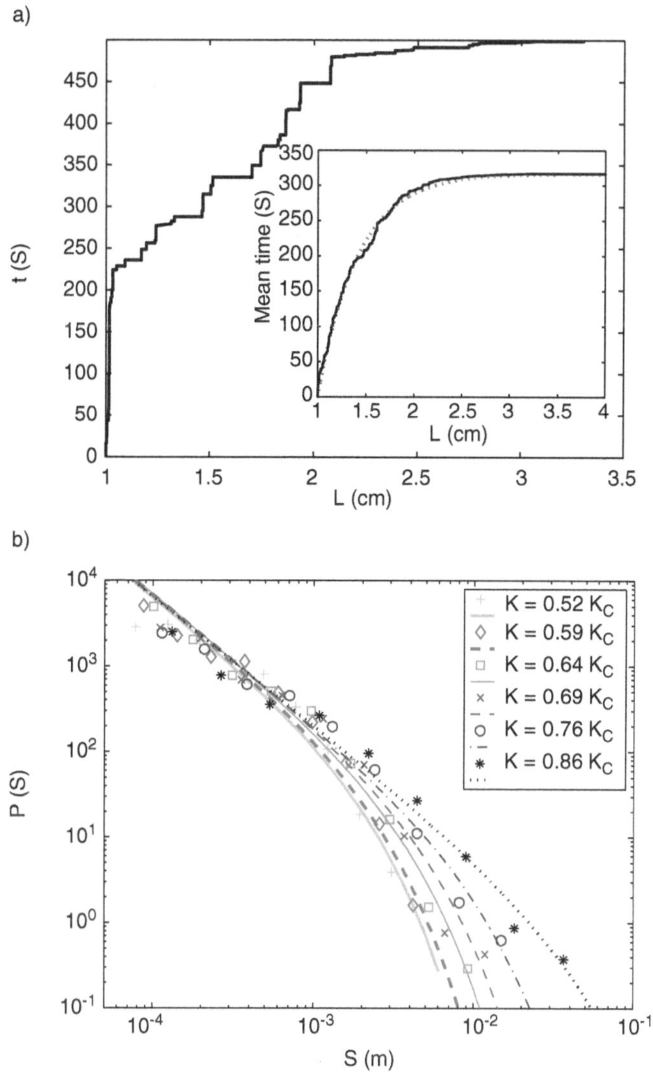

Figure 5: Experiment on the slow crack propagation on a sheet of paper. a) Time is plotted as function of the crack length for a single experiment. The plot shows crack jumps and crack arrest. In insert, average time to reach L (10 experiments with $F = 270$ N and $L_i = 1$ cm) and prediction for thermally activated growth from Eq.3 [7] (dotted curve). b) Step size distributions (symbols) for various values of the stress intensity factor. The continuous lines are best fits of experimental data done using the predictions of the 2D elastic model of thermally activated crack.

very good agreement with the functional form of Eq.3. The 2D elastic model of crack thermal activation predicts also the probability density function $p(s)$ of the crack jump sizes s. An example is given in Fig.5b), where $p(s)$ are plotted for various values of the stress intensity factor $K = F/(Wh)\sqrt{(L)}g(L/W)$. Here W

is the width of the sample, h its thickness and $g(L/W)$ takes into account the finite size effects. We see the good agreement between the measured PDF and the predicted ones. Finally the 2D model predicts for τ the following dependence on K_i (K at $L = L_i$)

$$\tau = \tau_0 \exp\left[-\frac{\alpha(K_c - K_i)^2}{2Y K_B T} \right] \tag{4}$$

where K_c is the toughness of the material, that is its critical stress intensity factor. The measured τ of thin sheets of paper seem to have the functional dependence on K_i of Eq.4, but at the moment the statistic on τ is not yet enough to make an accurate statement.

As a conclusion, we have shown that slow crack nucleation in heterogeneous materials presents, close to the failure time of the samples, several critical features. We have also seen that the real control parameter of this critical behavior is time and all the other parameters of the experiments (elastic properties of the materials, loading speed etc) are taken into account by the time of life τ. The functional behaviour of τ on the applied stress is predicted by activation models, which, in the case of experiments in thin sheets of papers, can predict also the statistics of crack jumps and the mean behaviour of the crack. However many problems are still open for understanding the role of activation processes in crack formation. The main one is the dependence on temperature of the life time. Indeed in all of the experiments the temperature estimated by the models is much higher than the real one.

We have shown that, in simple models, the disorder of the material strongly enhances the role of thermal fluctuations. However the correspondence of these theoretical results on real systems is still unclear. Furthermore there is another important open problem, specifically the different dependence of τ on the applied stress for systems close to 2D (Eq.4) and for 3D systems (Eq.2). Both equations seem to fit the experimental data. However one may wonder whether Eq.2 is an approximation of a much more complex behaviour. The final question concerns the role of viscoelasticity in the delayed crack. This role is still unclear in these systems. All these problems requires many new experiments in order to solve them. Work is in progress.

References

[1] Garcimartin A., Bellon L., Guarino A., Ciliberto S. Statistical analysis of crack precursors, Phys. Rev. Lett., 79, 3202 (1997).

[2] Guarino A., Garcimartin A., Ciliberto S. An Experimental Test of the Critical Behaviour of Fracture Precursors, Eur. Phys. J. B 6, 13 (1998).

[3] Guarino A., Garcimartin A., Ciliberto S. Failure time and microcrack nucleation, Europhysics Letters, 47, 456 (1999).
[4] Guarino A., Ciliberto S., Garcimartin A., Zei M., Scorretti R. Failure time and critical behaviour of fracture precursors in heterogeneous materials, European Physical Journal, B 26, 141–151 (2002).
[5] Guarino A., Scorretti R., Ciliberto S. The effect of disorder on the fracture nucleation process, Physica D 158, 83–104 (2001).
[6] Politi A., Ciliberto S., Scorretti R. Failure time in the fiber-bundle model with thermal noise and disorder, Phys. Rev. E. 66 (2), 026107 (2002).
[7] Santucci S., Vancl L., Guarino A., Scorretti R., Ciliberto S. Thermal activation of rupture and slow crack growth in a model of homogeneous brittle materials, Europhysics Letters Vol.62, 3, 320 (2003).
[8] Santucci S., Vanel L., Ciliberto S. Sub-critical statistics in rupture of fibrous materials: model and experiments, submitted.

Earthquakes and Acoustic Emission – Carpinteri & Lacidogna (eds)
© 2007 Taylor & Francis Group, London, ISBN 978-0-415-44402-6

RESPONSE OF FRACTURE NUCLEATION SITE TO WEAK MECHANICAL PULSES

V.S. Kuksenko & E.E. Damaskinskaya
Department of Strength Physics, Ioffe Physical Technical Institute Russian Academy of Sciences, St.Petersburg, 194021 Russia

ABSTRACT

The study is concerned with the influence of weak mechanical pulses on development of the fracture nucleation site. The macroscopic fracture of solids is considered as not an abrupt discrete event, but as a kinetic process. We assume that the fracture development in a heterogeneous material can be described in terms of the two-stage model of fracture. At the first stage, a disperse accumulation of noninteracting cracks occurs. When the local defect concentration in some area reaches the threshold value, the prerequisite for the accelerated defect formation is set up. This area is referred to as the fracture nucleation site. Further defect accumulation provides conditions for the loss of stability of the fracture nucleation site, as a result, a larger defect is formed. Experimental data obtained for a wide fracture scale (from microcracks to earthquakes) supported the validity of the two-stage model. A study of acoustics emission (AE) due to the cyclical deformation of samples of granite, marble and sandstone was made. It has been found that AE pattern apparently depends on the stress magnitude. At a low stress, during the unloading after the first loading cycle AE becomes negligible and at subsequent cycles increases insignificantly. At a high stress resulting in formation of a fracture nucleation site, the unloading is not followed by a sensible drop in the AE and succeeding cycles may bring about rise of AE intensity. This is in agreement with the fact that at a certain stage of formation the fracture nucleation site develops independently of the stress magnitude.

1 INTRODUCTION

Kinetic approach to the problem of strength of solids (Zhurkov [1]) and extensive experimental data on processes of crack generation and developing (Tamuzh [2], Zhurkov [3]) suggested a model of fracture of heterogeneous materials.

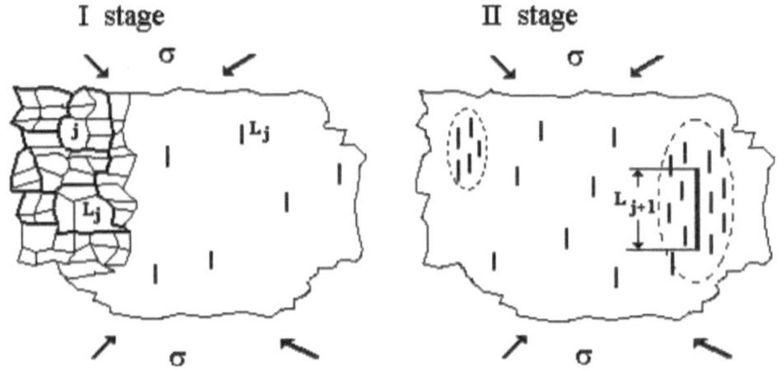

Figure 1: Scheme of the two-stage model of fracture of heterogeneous material.

2 PROBLEM SETTING

We assume that the fracture development in a heterogeneous material can be described in terms of the two-stage model of fracture (Kuksenko [4]). Briefly, the key points of the model can be described as follows.

Loading of a heterogeneous material results in the crack formation. At the first stage, a disperse (diffuse) accumulation of noninteracting cracks takes place (Fig.1). The sizes of these cracks are dictated by the structure scale level. The threshold value of the local defect (crack) concentration is reached in some area, the prerequisite for the accelerated defect formation is set up. This area is referred to as the fracture nucleation site. Further defect accumulation creates conditions for the loss of stability of the fracture nucleation site and, as a result, a defect is formed at the next level corresponding to the next characteristic size of the material structure.

The model has the advantage of validity at any structure scale level. It enables to extend the results of the laboratory sample tests to higher structure scale levels. In our earlier works (Kuksenko [4]) we formulated the statistical criterion for the fracture nucleation site formation. It was found that a simultaneous decrease in the mean temporal (Δt) intervals between chronologically successive defects and an increase in relevant variation coefficients ($V_{\Delta tr}$) indicates formation of a fracture nucleation site. It was shown (Myachkin [5]) that the bulk of AE in rocks results from the generation of cracks. It was found as well (Kuksenko, [6]) that parameters of cracks and related AE signals are correlated. So, examination of AE gives an insight into the fracture development.

Figure 2 shows the distribution of hypocenters of AE signals registered during loading of granite samples (Lockner [7], Tomilin [8]). The first (disperse) stage exhibits the uniform distribution of the AE sources through the bulk of a sample. At this stage, values of mean temporal intervals vary only slightly and variation coefficients is close to 1. From time T1, a decrease of Δt and increase of $V_{\Delta t}$ are observed. Localization of the defect formation begins. This is illustrated in Fig.2b, which shows distribution of hypocenters of AE sources registered after T1. The fracture switches to the second nucleation stage.

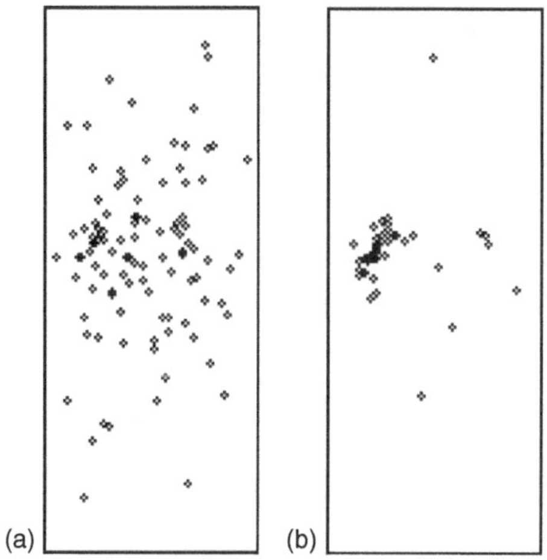

Figure 2: Spatial distribution of hypocenters of AE signals.

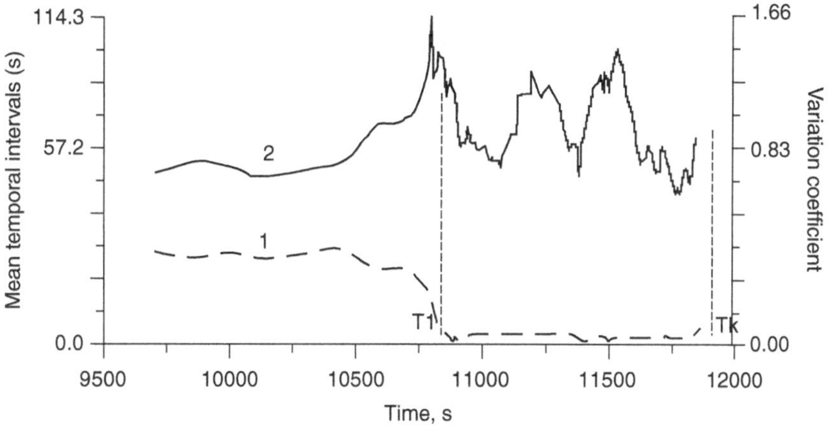

Figure 3: Variations of statistical parameters at the analysis of AE signals: mean temporal intervals (1), variation coefficient of mean temporal intervals (2).

This method for identification of the nucleation stage onset has been extensively and successfully applied to predict the time and place of catastrophic changes in a material, however, it has some notable drawbacks. For instance, despite improvements in the AE registration techniques, the problems of exact location (deriving the coordinates of hypocenters of AE signals) and gaining adequate statistics are not conclusively solved. Besides, this method implies monitoring of the fracture development, starting with earlier stages. However, this cannot be done when testing large objects and objects that has been in use for a long time. In such cases the tests involve considerable loading resulting in not only premature aging but, perhaps, in damage to the object.

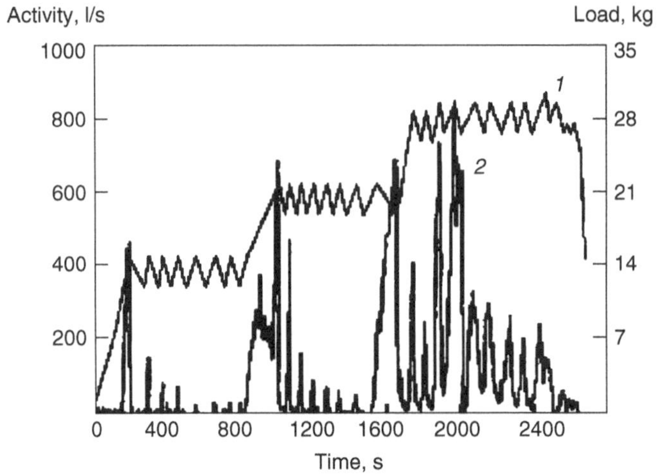

Figure 4: Temporal dependencies of load (1) and AE activity (2).

This work was aimed at investigation of behavior of the fracture nucleation site at various conditions of loading. It was expected that the regularities to be found would permit the employment of weak loading to evaluation of the extent to which a fracture nucleation site is ready for transformation to a larger defect.

3 EXPERIMENTAL DATA AND ANALYSIS

Granite samples shaped as a rectangular prism of the $140 \times 70 \times 70\,\text{mm}^3$ size were subjected to loading with various magnitudes and patterns. During testing, the AE signals were registered by a fast ten-channel devise. Analysis of flow of AE signals and calculation of coordinates of the AE source hypocenters permitted us to determine with fair accuracy the time and place of arising of a fracture nucleation site capable of disintegration of a specimen. Fig. 4. shows the variation of load with time. In the same figure changes of the AE activity are given. There were three exposures of a sample to different magnitudes of the loading.

It is seen from the plots, that in the first test AE intensity decreased rapidly, with increasing cycle number, down to the background noise. During the second test, the AE pattern was similar. Spatial distribution of hypocenters of AE sources implies the uniform accumulation of defects over the sample's bulk. The third test resulted in triggering of the process to the (nucleation) stage. At that moment the static loading was interrupted and the sample was subjected to the cycled loading with the magnitude equal to that in previous tests. In this test, the AE response changed dramatically (Fig. 4).

Activity behavior was analyzed separately for increasing and decreasing load in three series of loading to find correlation between AE activity and increasing/decreasing circles of load. Normalized graphs of averaged AE activity are

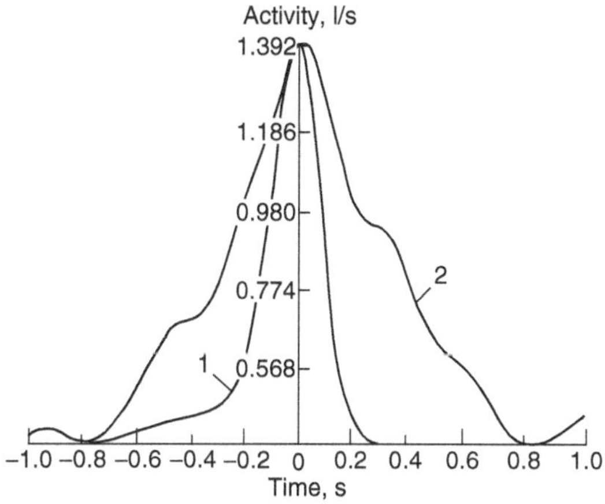

Figure 5: Temporal dependencies of normalized AE activity for disperse (1) and nucleation (2) stage.

represented for first (curve 1) and third (curve 2) series of loading at Fig.5. The negative time semiaxis corresponds to increase of load, the positive one – to decrease. The curve 1 shows that, during the first test, the acoustic activity is largely concentrated within narrow time interval close to the moment of the maximal load. Besides, the AE response is markedly asymmetrical due to rapid drop in AE intensity on unloading. The integrated activity at loading is approximately twice as large as that at unloading. The time distribution of AE response is considerably wider at the nucleating stage (Fig.5, curve 2). The absence of asymmetry in AE response is noticeable as well: ratio between integrated activities at loading and unloading is close to 1.

4 CONCLUSIONS

The major implication of this study is that the condition of a heterogeneous material containing the mature fracture nucleation site notably differs from the initial condition. In addition, we conclude that:

(1) Arising of a fracture nucleating site in a material considerably affects the AE response of the sample to variations of load.
(2) Analysis of the AE response to the cycled loading permits identification of the fracture stage (disperse or nucleation).
(3) Development of a fracture nucleation site may be stimulated by a weak cycled loading.
(4) Validity of the model at any structure scale level implies that methods described above can be applied to investigation of nucleation of earthquakes.

ACKNOWLEDGEMENTS

This work was supported by Russian Foundation for Basic Research (Grant N 02-05-08003, N 02-05-39017, N 03-05-64831, N 04-05-65287).

References

[1] Zhurkov S.N., Kinetic concept of the strength of solids, Int. J. Fracture Mech. 1, pp. 311–323, 1965.
[2] Tamuzh, V.P., and Kuksenko, V.S., Fracture micromechanics of polymer materials (Martinus Nighoff Pub., The Hague, Boston, London), 1981.
[3] Zhurkov, S.N., Kuksenko, V.S., Petrov, V.A., Principles of the kinetic approach of fracture prediction, Theoretical and Applied Fracture Mechanics, 1, pp. 271–274, 1984.
[4] Kuksenko V., Tomilin N., Damaskinskaya E., and Lockner D., A Two-stage Model of Fracture of Rocks // Pure Appl. Geophys. 146 (2). pp. 253–263, 1996.
[5] Myachkin V.I., Kostrov B.V., Sobolev G.A., Shamina O.G., Laboratory and theoretical investigations of process of Earthquake forming, Izvestiya Academy of Sciences, Earth Physics, N 10, pp. 2526–2530, 1974.
[6] Kuksenko V.S., Lyashkov A.I., Mirzoev K.M. et al, Correlation between crack size at loading and duration of elastic energy release, Doklady Academy of Sciences, v.264, N4, pp. 846–848, 1982.
[7] Lockner D.A., Byerlee J.D., Kuksenko V., Ponomarev A., and Sidorin, A., Observations of quasistatic fault growth from acoustic emissions, In Fault Mechanics and Transport Properties of Rocks (ed. Evans, B. and Wong, T.-F.) (Academic Press, London) pp. 3–31, 1992.
[8] Tomilin N.G., Damaskinskaya E.E., Kuksenko V.S., Formation of a fracture focus during the deformation of heterogeneous materials (granite), Phys. Solid State, v.36, N10, pp. 1649–1653, 1994.

Earthquakes and Acoustic Emission – Carpinteri & Lacidogna (eds)
© 2007 Taylor & Francis Group, London, ISBN 978-0-415-44402-6

FRACTURE IN TWO DIMENSIONS (IN PAPER): ACOUSTIC EMISSION STUDIES AND THEORETICAL LESSONS

L.I. Salminen, J. Rosti, J.M. Pulakka, & M.J. Alava
Laboratory of Physics, Helsinki University of Technology, Finland

ABSTRACT

Acoustic emission experiments in paper, a (quasi)-two dimensional disordered material, produce evidence of criticality, in the language of statistical physics. The energies of events follow a power-law, scale-free probability distribution. Two basic setups, usual mode I tensile tests and a nip-in-peel one have been studied. These are compared with each other, and with the pertinent theories of slow fracture and crack line propagation.

1 INTRODUCTION

Acoustic emission (AE) is one of the statistical phenomena in fracture, that shows signs of universal phenomena, such as are encountered in many disguises in statistical physics (Lockner, Petri, Krysac, Guarino [1, 2]). The essential idea is that transient elastic waves are detected, once they get generated through the release of elastic energy due to microcracking and (main) crack propagation. The time-series at a single AE detector can be described by "crackling noise": silent intervals are separated by AE events of varying length and amplitude/energy (Bouchaud [3]). Numerous studies have elucidated the statistical laws that describe AE. In general, these relate to usual mode I or mode III-type loading conditions, and the common feature is that they exhibit *scalefree* features. The probability distribution function (pdf) of event energies is most often of power-law type, $P(E) \sim E^{\beta}$ with $\beta = 1 \ldots 2$, and e.g. the event intervals are most often, similarly, found to obey such fat-tailed pdf's.

Other "fractal" characteristics relate to the properties of final crack surfaces. It has been discovered that both for 2d and 3d geometries fracture interfaces can be described by self-affine fractals (Bouchaud [4]). Thus e.g. the mean-square fluctuations w increase with the scale of observation as $w \sim l^{\chi}$, with the roughness exponent χ depending on the range of observation and dimensionality, among others. The reasons for scale-invariance are still being debated

(Bouchaud, Bouchaud et al., Räisänen, Hansen and others [4,5,6,7]). In this context, a very clean scenario is presented by the propagation of an interfacial crack (line) between two three-dimensional plates, of e.g. plexiglass. Now it is clear that if the crack front exhibits non-trivial fluctuations these could perhaps be accounted for by a dynamical model for such a line, in the presence of the right kind of disorder, and with the right kind of interactions (Schmittbuhl, Le Doussal, Rosso; Schmittbuhl, Alava [8,9]). However, where as renormalization group calculations and numerical models indicate $\zeta \sim 1/3 \ldots 0.38$, the experimental evidence points towards a much more complicated scenario (Schmittbuhl, Delaplace [10]). The roughness exponent is higher (upto 0.6), and there is clear evidence of the presence of localized activity, weak oscillations, or avalanches (Måløy [11]). In general, theoretical models for statistical fracture as found in the literature do not include more complicated effects than locally varying material properties (strength, elastic modulus) and simplified load-sharing. They do imply the presence of scale-free dynamics, and in the most simple case of democratic (global) load-sharing models (fiber bundle ones) (Kloster [12]), these properties can be defined and measured exactly. In more elaborate systems, like random fuse networks (Herrmann [13]) which are a scalar analogy of quasi-brittle fracture, these features still persist (e.g. in terms of a broad $P(E)$ distribution, and the notion of avalanches) (Zapperi [14]).

Here we consider paper as a test case of theories of fracture in the presence of structural randomness. This is of interest since paper is (almost) two-dimensional, among others (Kertész [15]). In the following, we concentrate on two kinds of experiments (Salminen, 2002; Salminen et al. [16,17]). The more recent is based on a "paper peeling", in which a crack line is forced to propagate along the sheet plane, thus separating the sheet into two halves. In addition, we consider the usual mode I fracture, and highlight the scaling by using the strain rate as an extra control parameter.

2 EXPERIMENTAL SETUP

Normal newsprint paper samples (size 100 mm by 100 mm) and laboratory sheets (size 70 mm by 15 mm) were tested in two geometries, in mode I (tensile) and peel-in-nip. The later produces very large fracture surfaces. Due to the lack of constraints the samples could have out-of-plane deformations in tensile tests, too, and none of the three fracture modes (I, II, III) is excluded on the microscopic level. The deformation rates $\dot{\varepsilon}$ varied between 0.1%/min and 100%/min. During the experiment we acquire bi-polar acoustic amplitudes simultaneously by piezoelectric sensors, as a function of time. In most cases the AE apparatus consisted of two transducers, +27 dB amplifiers and continuous 12-bit data-acquisition. The time-resolution of the measurements was 2.5 μs and the data-acquisition free of deadtime. We made 20 identical repetitions for statistics. The strain rates are such that the sound velocity is much faster than the timescales implied by $\dot{\varepsilon}$. The acoustic time-series are reformed offline by thresholding, detection of continuous and coherent oscillatory events, and the calculation of event energy E, the sum of squared amplitudes within the event. Events are separated by silent (i.e. amplitude below threshold level) waiting intervals t. In general the energy of the event is expected to be proportional both to the damaged area and to the stress in that area.

3 TENSILE FAILURE

Here, one can compare the results with mean-field-like avalanche models (fiber bundles), that would imply using scaling $P(E) \sim E^{-(5/2+1)/2} = E^{-7/4}$, and simple simulations (Salminen 2002, Minozzi [16,18]). One particularly interesting twist is demonstrated in Fig. 1, in which the accumulation of AE energy is divided according to whether it has originated, in a strain-controlled experiment, before or after the maximum stress σ_c. Excluding the last data point (for 100 mm/min) the typical total AE energy does not vary much, but the fractions do very much so. Since, the corresponding energy scalings imply in general $\beta \sim 1.2 \pm 0.1$, we must conclude that the microscopic crack growth dynamics is independent of the "macroscopic failure point", or whether catastrophic failure happens right at σ_c or beyond that. It is to be emphasized that in this respect it should not be of importance whether a test is made in stress- or strain-controlled circumstances. The acoustic emission energy accumulates close to the failure point in an exponential fashion, as depicted in Fig. 2. There is evidence of a slow increase of a background AE level, which may be sensitive to particular experimental choices (thresholding etc.). and prior to the maximum stress the accumulation is fast. Attempts to match the AE integral with a power-law-like divergence ($\int E_{AE}\, dt \sim (\sigma_c - \sigma(t))^{-a}$) are unsuccessful, and note that the data exhibits several decades of exponential growth. We may therefore conclude, that

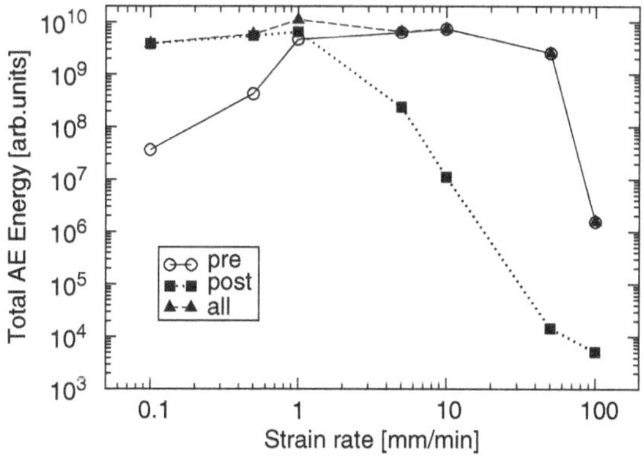

Figure 1: The fractions of AE energy, prior and after to the σ_c of the stress-strain curve, as a function of the strain rate in mode I experiments.

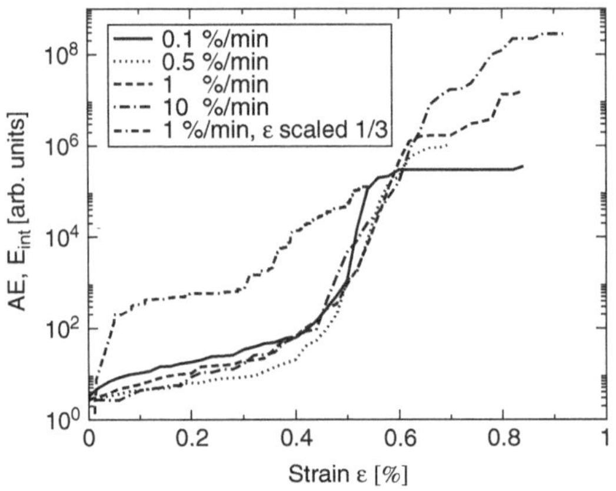

Figure 2: Integrated AE energy vs. strain, for four data sets with the same paper, but with different strain rates, and a fifth one as a comparison). Notice the exponential increase of the AE energy accumulation. The rounding-off of the curves signals the regime in which individual samples fail.

there is no evidence of a "finite-time singularity" (Johansen, Shcherbakov, Guarino [19, 2]), in the fracture of paper.

4 CRACK LINE PROPAGATION IN A RANDOM ENVIRONMENT

In this case, the fracture dynamics presents steady-state conditions, unlike in a mode I experiment (say), and thus there are intriguing possiblities for the analysis of e.g. stationary AE timeseries. The first observation is that the dynamics is again scale-invariant (Salminen et al. [17]), and in both the AE energy and event-interval pdf's we see clear observations of fat-tailed, broad characteristics (Fig. 3). There are three main observations as mode I and in-plane fracture are compared: i) the β-exponent is much larger for the latter one. This may be attributed perhaps to the tendency, noted by various theoretical approaches, for the crack dynamics to be very much localized around the crack line on the expense of bulk damage ahead of it (Zapperi et al., Åström [20]). ii) A similar difference can be noticed in the waiting time statistics, as well, and, iii) most importantly the in-plane peel test shows clear evidence for an *intrinsic scale*. Other tests at varying strain rates imply that the associated t_i derives from a fundamental length, closely related to the size of individual fiber-to-fiber bonds (30–50 μm). These, or the fibers themselves, are the fundamental building blocks of the structure (Salminen [17]). Finally, given the steady-state conditions the AE time series can be used to infer different kinds of correlations in the energy release. Theoretically, this is partly uncharted territory; in statistical physics avalanche models the avalanches (AE events in our language) are often rather uncorrelated, but can also be described by a broad pdf (eg. for the size or duration). In Fig. 4 we consider the autocorrelation

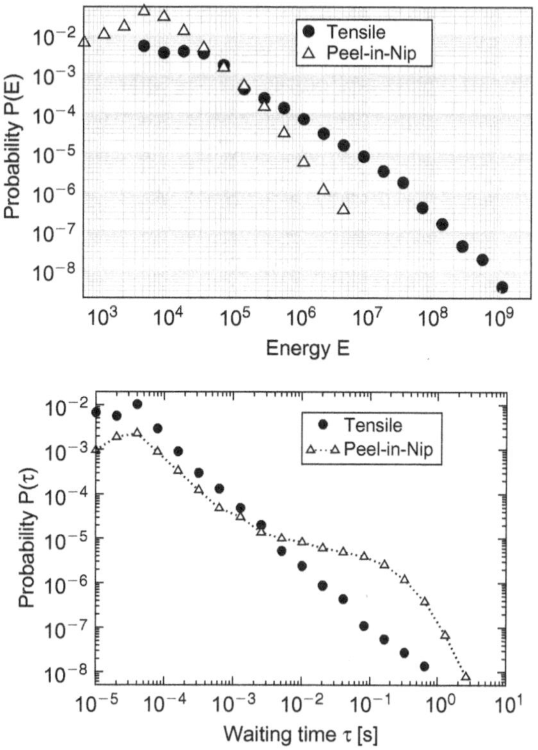

Figure 3: Top panel: the energy distribution $P(E)$ for two experimental setups. Both exhibit *power-law* statistics, but with different exponents, and that the mode I has a much smaller exponent (close to 1.2). Lower panel: The distribution of time intervals t for the mode I and in-plane experiments. Both obey an Omori's law -like scaling, and the mode I one scales close to $P(\tau) \sim \tau^{-1}$. For the latter one there is a distinct time-scale (see text).

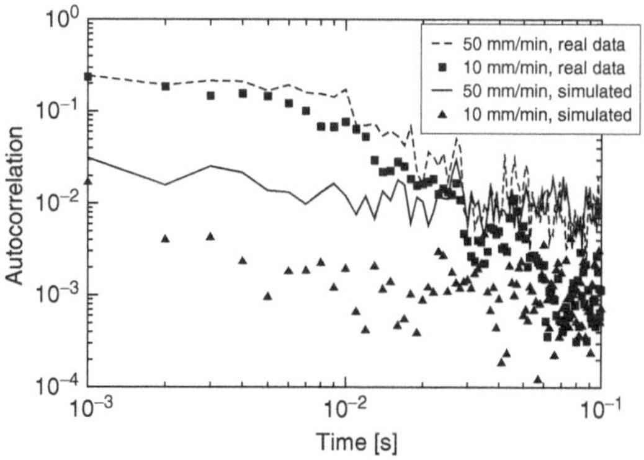

Figure 4: The autocorrelation function of released energy from the in-plane experiment, for two strain rates (10 and 50 mm/min). The data is compared to two artificial time series with the same interval and energy statistics.

function of the AE energy signal, after thresholding and dividing into discrete 1 ms timebins. The data is compared to two test sets, obtained by randomly rearranging the sequence of events and empty intervals. While the noise levels of the two strain rates are different (the average AE energy per bin increases with the rate), some conclusions can be drawn. The AE release is correlated upto a timescale that is only weakly dependent on the strain rate, up to about 0.02 …0.08 s.

5 CONCLUSIONS

In this work, we have searched for universality in the fracture of paper, as both a test of current theories of statistical fracture and to present some landmarks for further theoretical work. In both the cases studied clear evidence is found of broad energy distributions. The two main conclusions have been: the mode I one is far smaller than that resulting from any model as fuse networks or fiber bundle models; and, in the peeling test a β-exponent is recovered, such that there is again no known theoretical framework. Meanwhile, the temporal statistics implies scale-invariance if measured for instance in terms of the event interval statistics. Both mode I and in-plane fracture, using the same material, imply correlations in the fracture process. In the development of damage prior to maximum stress in mode I we observe exponential growth of released energy. This is similar to random damage, or softening due to reduced elastic modulus, which leads to a localized fracture zone in quasi-brittle materials (ice, concrete etc.) (Van, Delaplace [21]). Unfortunately we do not have enough statistics to use "b"-analysis, the possible variation of the β-exponent along the stress-strain curve. This would also be a stringent test in the statistical physics sense, since many models imply that while the β-exponent would not change, the cut-off of the pdf would, thus resulting in a net β-exponent from the combination of these two. On the other hand, the data is clearly in contrast to attempts to describe the approach of σ_c in terms of the concept of a "critical point". In the stationary statistics correlations can be seen, but it is unclear whether these result from structural correlations – the line getting "pinned" by stronger interfacial regions – or its intrinsic dynamical response. To summarize, paper fracture allows to test many simple scenarios of statistical fracture, and still more work could be done e.g. to consider the detailed properties or shape of the AE events in the various scenarios. A further prospect is to connect AE and damage dynamics (localization) to the roughness of the final crack line.

ACKNOWLEDGEMENTS

We would like to acknowledge A. I. Tolvanen and K.J. Niskanen (KCL Ltd), as well as the Center of Excellence program, Academy of Finland, for support.

References

[1] Lockner, D.A., et al., Nature 350, 39 (1991); Petri, A., et al., Phys. Rev. Lett 73, 3423 (1994); Krysac L.C. and Maynard, J.D., Phys. Rev. Lett. 81, 4428 (1998).
[2] Guarino, A., Garcimartin, A., and Ciliberto, S., Eur. Phys. J. B 6, 13 (1998); Garcimartin A., et al., Phys. Rev. Lett 79, 3202 (1997).
[3] Sethna, J.P., Dahmen, K.A., and Myers, C.R., Nature (London) 410, 242 (2001).
[4] Bouchaud, E., J. Phys. Cond. Mat. 9, 4319 (1997).
[5] Bouchaud, E., Bouchaud, J.P., Fisher, D.S., et al., J. Mech. Phys. Solids 50, 1703 (2002).
[6] Räisänen, V. I., Alava, M.J., Seppälä, ET., and Duxbury, P.M., Phys. Rev. Lett. 80, 329 (1998).
[7] Hansen, A. and Schmittbuhl, J., Phys. Rev. Lett. 90, 045504 (2003); Nukala, P. K. V. V., Simunovic, S. and Zapperi, S., cond-mat/0311284; Reurings F. and Alava, M.J., cond-mat/0401592.
[8] Schmittbuhl, J., Roux, S., Vilotte, J.P., and Måløy, K.J., Phys. Rev. Lett. 74, 1787 (1995); Le Doussal, P.,Wiese, K.J., and Chauve, P. Phys. Rev. E66, 174201 (2002); Rosso A. and Krauth W. Phys. Rev. E65, 025101 (2002).
[9] Schmittbuhl, J., Hansen, A., and Batrouni, G.G. Phys. Rev. Lett. 90, 045505 (2003); Alava, M.J., and Zapperi, S., Phys. Rev. Lett. 92, 049601 (2004).
[10] Schmittbuhl, J., and Måløy, K.J., Phys. Rev. Lett. 78, 3888 (1997); Delaplace, A., Schmittbuhl, J., and Måløy, K.J., Phys. Rev. E 60, 1337 (1999).
[11] Måløy, K.J. and Schmittbuhl J. Phys. Rev. Lett. 87, 105502 (2001).
[12] Kloster, M., Hansen, A., and Hemmer, P.C., Phys. Rev. E56, 2615 (1997).
[13] Chapters 4–7 in *Statistical models for the fracture of disordered media*, ed. Herrmann, H. J. and Roux, S., (North-Holland, Amsterdam, 1990).
[14] Zapperi, S. et al., Phys. Rev. Lett. 78, 1408 (1997).
[15] Kertész, J., Horvath, V. K., and Weber, F., Fractals 1, 67 (1993).
[16] Salminen, L.I., Tolvanen, A.I., and Alava, M.J., Phys. Rev. Lett. 89, 185503 (2002).
[17] Salminen, L.I. et al., submitted for publication.
[18] Minozzi, M., Caldarelli, G., Pietronero, L., and Zapperi, S., Eur. Phys. J. B36, 203 (2003).
[19] Johansen, A. and Sornette D., Eur. Phys. J.B 18 163 (2000); Shcherbakov R., and Turcotte D.L., Theor. Appl. Fract. Mech. 39 245 (2003).
[20] Zapperi S., Herrmann H.J., and Roux S. Eur. Phys. J.B 17, 131 (2000); Åström J.A., Alava M.J., and Timonen J., Phys. Rev. E62, 2878(2000).
[21] Van P., Papenfuss, C., and Muschik, E., Phys. Rev. E62, 6206 (2000); Delaplace, A., Pijaudier-Cabout, G., and Roux, S., J. Mech. Phys. Solids 44, 99 (1996).

2.2 AE testing on concrete and concrete structures

Earthquakes and Acoustic Emission – Carpinteri & Lacidogna (eds)
© 2007 Taylor & Francis Group, London, ISBN 978-0-415-44402-6

MICROMECHANICS OF CORROSION CRACKING IN REINFORCED CONCRETE BY AE

F. Uddin A.K.M. & M. Ohtsu
Graduate School of Science and Technology, Kumamoto University,
Kumamoto 860-8555, Japan

ABSTRACT

Nucleation of micro-cracking due to corrosion is detected and analyzed by AE-SiGMA procedure, by which crack kinematics of locations, types and orientations are quantitatively determined. Since these kinematical outcomes are obtained as three-dimensional (3-D) locations and vectors, 3-D visualization is developed by using VRML (Virtual Reality Modeling Language). The procedure is applied to corrosion cracking. Kinematics of micro-cracking due to corrosion are studied by numerical analysis BEM. Thus, relations between the stress intensity factors and cracking modes are investigated for different crack-patterns. It is confirmed that micromechanics of corrosion cracking are dominantly of mode-I failure associate with mixed-mode and mode-II.

1 INTRODUCTION

Corrosion of reinforcement in concrete is a major deterioration in reinforced concrete. For making decision on maintenance and repair, identification of cracking mechanisms due to corrosion is significantly important. Expansion caused by corrosion product generates micro-cracking, of which mechanisms can be investigated experimentally by acoustic emission (AE). AE-SiGMA procedure is developed as a powerful technique for moment tensor analysis of AE (Ohtsu [1]). Crack kinematics of locations, types and orientations are quantitatively determined (Ohtsu [2]). Because these kinematical outcomes are obtained as three-dimensional (3-D) locations and vectors, virtual reality modelling language (VRML) is applied. Thus, failure processes of corrosion cracking are clarified by using 3-D visualisation of AE-SiGMA procedure. To investigate mechanisms of micro-cracking, numerical analysis by the boundary element method (BEM) is conducted on the basis of linear elastic fracture mechanics (LEFM). The maximum circumferential stress criterion based on LEFM was successfully applied to trace the crack extension (Farid and Ohtsu [3]). By employing the two-domain BEM, cracking mechanisms are evaluated from the stress intensity factors.

2 EXPERIMENT

To study crack patterns due to rebar corrosion in concrete, experiments were carried out. Concrete was made up of mixture as water (W):cement (C):sand (S):gravel (G) = 0.5:1.0:2.41:2.95 by weight. The maximum size of aggregate was 20 mm. Slump value and air content were 7.0 cm and 5.0%, respectively. The dimensions of concrete specimens were 25 cm × 25 cm × 10 cm. Two concrete specimens were casted with cover thickness 4 cm. One is named specimen-A and other is specimen-B. To investigate other types of cracks than the surface crack, a pre-cracked notch with 1 mm width was created from the nearest side to the hole in specimen-B. This is because the surface crack is always nucleated prior to other cracks. To simulate corrosion cracking, an expansion test was conducted. In Figures 1 and 2, a circle of 3 cm diameter represents the location

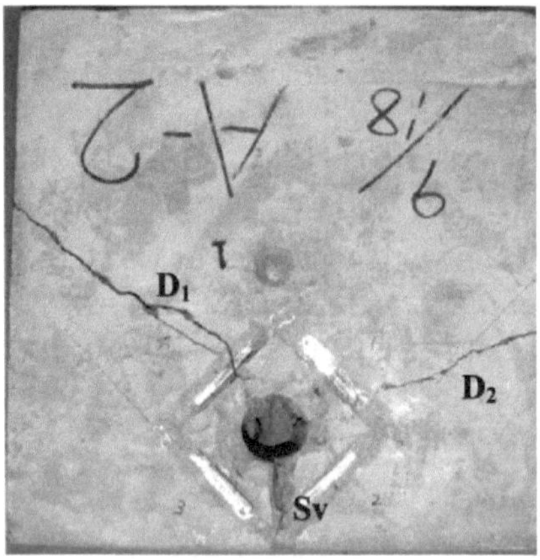

Figure 1: Observed cracks in specimen-A.

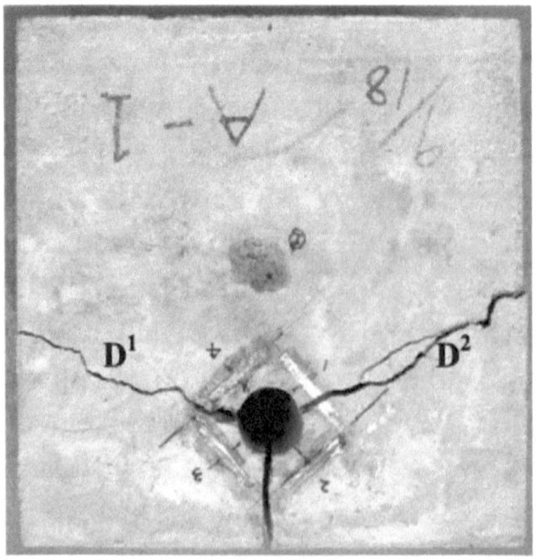

Figure 2: Observed cracks in specimen-B.

of the reinforcement where the expansion pressure is applied by dolomite paste and the expansion pressure was measured by using a pressure gauge embedded. Observed cracks after the test are found in the figures. Mechanical properties were obtained from cylindrical specimens at the age of 28 days. Compressive strength was 37.9 MPa. The modulus of elasticity and Poisson's ratio were 29.7 GPa and 0.22, respectively, which were applied to in BEM analysis.

3 CRACK KINEMATICS BY AE-SiGMA

By taking into account only P wave motion of the far-field term of Green's function in an infinite space, displacement Ui(x,t) of the first motion is obtained,

$$Ui(x,t) = -1/(4\pi\rho\upsilon_p{}^3)(r_i r_p r_q/R)(dS(t)/dt)M_{pq}. \quad (1)$$

Here ρ is the density of the material and υ_p is the velocity of P wave. R is the distance between the source y and the observation point x, of which direction cosine is $r = (r_1, r_2, r_3)$. S(t) is the source-time function of crack motion. M_{pq} is the moment tensor. Considering the effect of reflection at the surface and neglecting the source-time function, amplitude A(x) of the first motion is represented,

$$A(x) = Cs(1/R)Ref(\mathbf{s},\mathbf{r})r_p M_{pq} r_q, \quad (2)$$

where Cs is the calibration coefficient including material constants in Eqn (1). **s** is the direction of the sensor sensitivity, and Ref(**s**,**r**) is the reflection coefficient. Since the moment tensor M_{pq} is symmetric, the number of independent components is six. Thus, multi-channel observation of the first motions at more than six channels is required to determine the moment tensor components by AE-SiGMA procedure. In the case of an isotropic material, the moment tensor M_{pq}, is defined as,

$$M_{pq} = (\lambda l_k n_l \delta_{pq} + \mu l_p n_q + \mu l_q n_p)\Delta V \quad (3)$$

where λ and μ are Lame's elastic constants. **l** is the unit direction vector and **n** is the unit normal vector to the crack surface as shown in Figure 3. ΔV is the crack volume. The classification of a crack is performed by the eigenvalue analysis of the moment tensor. Setting the ratio of the maximum shear contribution as X, three eigenvalues of the shear crack become X, 0, $-$X. Likewise, the ratio of the maximum deviatoric tensile component is set as Y and the isotropic tensile as Z. It is assumed that the principal axes of the shear crack are identical to those of the tensile crack. Then, the eigenvalues of the moment tensor for a general case are represented by the combination of the shear crack and

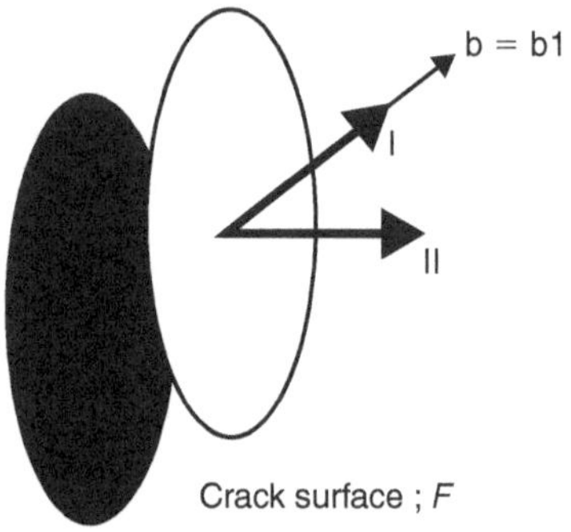

Figure 3: Modelling of a micro-crack.

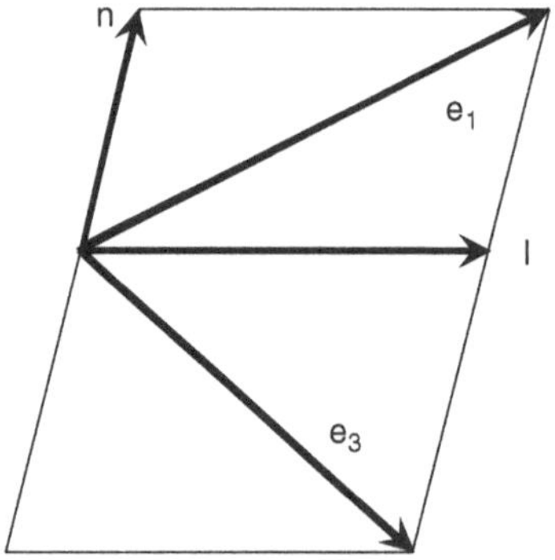

Figure 4: Eigenvectors by moment tensor analysis.

the tensile crack. Because relative values are determined in AE-SiGMA procedure, three eigenvalues E_1, E_2 and E_3 are normalized and decomposed as,

$$1.0 = X + Y + Z, \quad E_2/E_1 = 0 - 0.5Y + Z, \quad \text{and} \quad E_3/E_1 = -X - 0.5Y + Z, \quad (4)$$

where X, Y, and Z denote the shear ratio, the deviatoric tensile ratio, and the isotropic tensile ratio, respectively. In the present analysis, AE sources of which the shear ratios $X < 0.4$ are classified into tensile cracks. Sources of $X > 0.6$ are classified into shear cracks. Sources, of which X ratios are between 40% and 60%, are referred to as the mixed mode. Figure 4 shows the geometry among the eigenvectors $\mathbf{e}_1$ and $\mathbf{e}_3$, normal to crack surface $\mathbf{n}$ and crack motion direction $\mathbf{l}$. Three eigenvectors are $\mathbf{e}_1 = \mathbf{l} + \mathbf{n}$, $\mathbf{e}_2 = \mathbf{l} \times \mathbf{n}$, and $\mathbf{e}_3 = \mathbf{l} - \mathbf{n}$. Vectors $\mathbf{l}$ and $\mathbf{n}$ can be recovered from the following relations,

$$\begin{aligned} \mathbf{l} &= [(2 + 2l_k n_k)^{1/2}\mathbf{e}_1 + (2 - 2l_k n_k)^{1/2}\mathbf{e}_3]/2, \\ \mathbf{n} &= [(2 + 2l_k n_k)^{1/2}\,\mathbf{e}_1 - (2 - 2l_k n_k)^{1/2}\mathbf{e}_3]/2. \end{aligned} \quad (5)$$

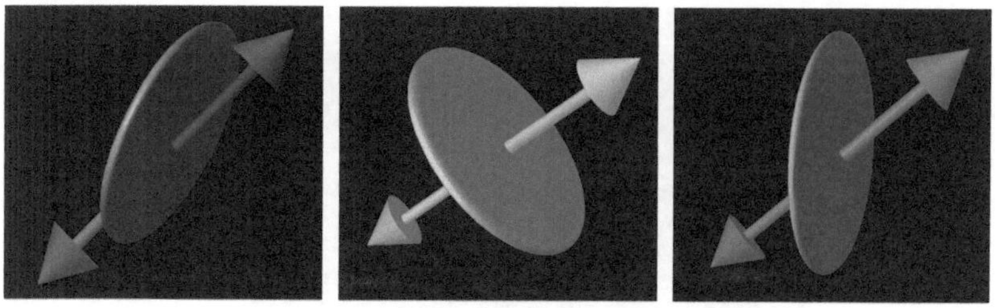

Shear crack Tensile crack Mixed-mode crack

Figure 5: VRML models for tensile, shear, and mixed-mode cracks.

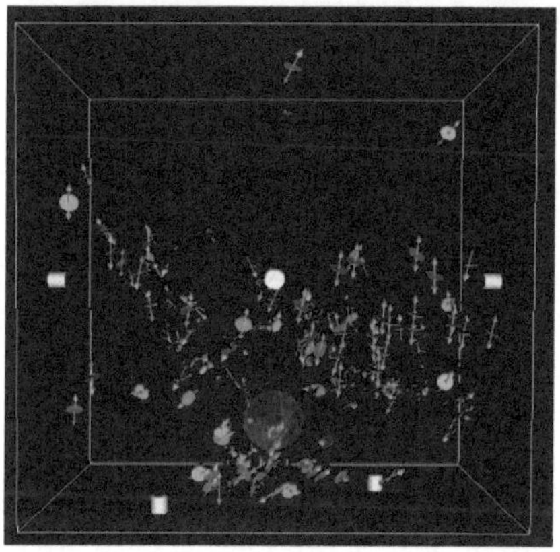

Figure 6: AE-SiGMA analysis visualised results of the specimen-A.

4 VISUALISATION OF MICRO-CRACKING

By conventional AE-SiGMA procedure, classification of cracks is made by symbols, whereas crack orientation is not easily recognized as 2-D projection. In this respect, VRML is introduced, as crack models of tensile, mixed-mode and shear are given in Figure 5. For specimen-A, 3-D visualization of AE-SiGMA procedure is shown in Figure 6. Shear cracks, tensile cracks and mixed-mode cracks are intensely observed near concrete cover, corresponding to the surface crack. The surface crack Sv, the diagonal cracks D_1 and D_2 start to initiate and extend. It is found that the surface cracks Sv is first nucleated and then the diagonal cracks D_1 and D_2 follow. 3-D visualization of AE-SiGMA procedure of specimen-B is shown in Figure 7. Shear cracks, tensile cracks and mixed-mode cracks are intensely observed corresponding to the diagonal cracks D^1 and D^2 in Figure 2 because of a pre-cracked notch is existed. Cracks distribute widely, corresponding to completion of the diagonal cracks.

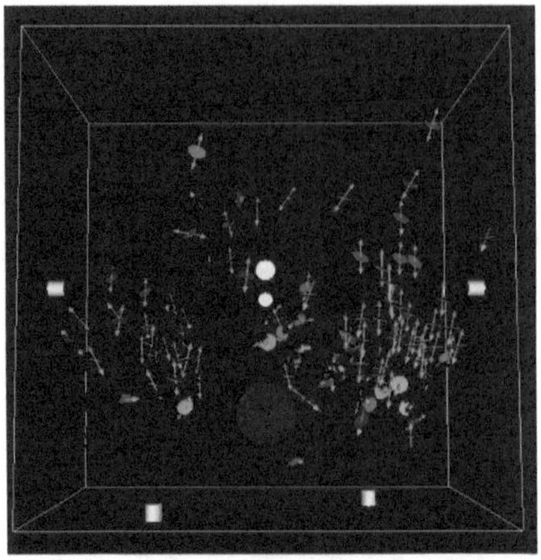

Figure 7: AE-SiGMA analysis visualised results of the specimen-B.

5 ANALYTICALLY MODEL

Traces of the surface crack and the diagonal cracks are analyzed by BEM in a similar manner to the previous research (Farid and Ohtsu [3]). Initial two-domain elements are stitched at the interface jointing the initial crack tip with the final point. All boundary meshes are of 5 mm long. Because expansive agent was employed in the experiment for producing expansive pressure. In BEM analysis, three types of expansive pressure are taken into consideration to simulate the expansion of corrosive products. These are uniform pressure, horizontal pressure and vertical pressure. The stress intensity factors K_I (mode I) and K_{II} (mode II) are determined from relative displacements at the crack-tip elements by Smith's one-point formula. The direction of the maximum tangential stress θ is determined from the maximum circumferential stress by Erdogan-Sih criterion (Erdogan and Sih [4]).

6 IDENTIFIED MICROMECHANICS FOR CORROSION CRACKING

Results analyzed by AE-SiGMA procedure are compared with those of BEM analysis. The dimensionless stress intensity factors K_I/K_{IC} and K_{II}/K_{IC} (Carpinteri [5]) are calculated at each step by substituting angles of micro-cracking from BEM analysis and from the experiment.

6.1 Surface crack Sv in specimen-A

The surface crack Sv is analyzed by BEM and compared with the actual crack trace in experiment. Through all the cracking steps, crack trace due to vertical

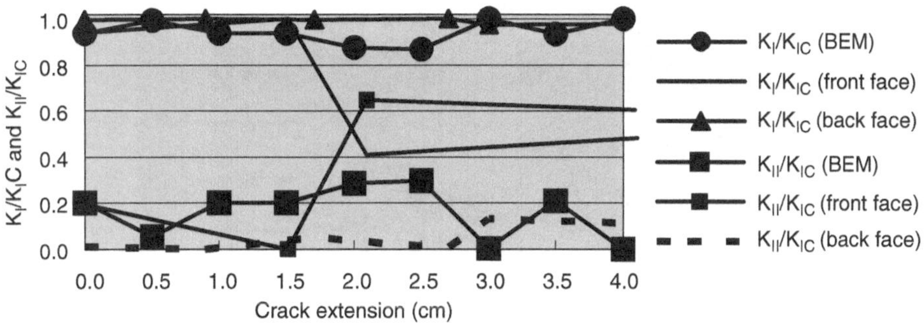

Figure 8: K_I/K_{IC} and K_{II}/K_{IC} during micro-cracking for surface crack Sc in specimen-A.

pressure is the closest to the actual trace in the experiment out of any other pressure distributions. K_I/K_{IC} and K_{II}/K_{IC} are calculated as shown in Figure 8. In the beginning of crack extension, K_I/K_{IC} is nearly equal to 1.0, so the pure mode I fracture occurs. K_{II}/K_{IC} obtained form the crack trace at front face increases gradually and in the middle stage of crack extension K_{II}/K_{IC} is greater than K_I/K_{IC}, where shear cracks could occur. Active nucleation of shear cracks is found in the results of AE-SiGMA procedure in Figure 6. It is concluded that for micro-cracking of the surface crack, tensile cracks eventually dominate shear cracks.

6.2 Diagonal crack D_1 in specimen-A

The diagonal crack D_1 is analyzed, simulating the crack extension after the surface crack. Crack trace due to uniform pressure is the closest to the actual traces K_I/K_{IC} and K_{II}/K_{IC} are calculated as shown in Figure 9. In only the beginning of crack extension, K_I/K_{IC} is nearly equal to 1.0. Then, K_{II}/K_{IC} increases gradually and in the middle stage K_{II}/K_{IC} becomes almost equal to K_I/K_{IC}. Also, K_I/K_{IC} values are mostly lower than those of the surface crack. As confirmed by 3-D VRML visualization of AE-SiGMA procedure in Figure 6, shear and mixed-mode cracks are intensely generated. It is concluded that for micro-cracking of the diagonal crack D_1, tensile cracks not only are dominant but also shear and mixed-mode cracks are active.

6.3 Diagonal crack D_2 in specimen-A

The diagonal crack D_2 is analyzed through all the cracking steps, crack trace due to uniform pressure is the closest to the actual traces for the both the front face and the back face of the crack. Analyzed K_I/K_{IC} and K_{II}/K_{IC} are shown in Figure 10. In the beginning of crack extension, K_I/K_{IC} is again nearly equal to 1.0. The values of K_I/K_{IC} are larger than those of the diagonal crack D_1, but the values of K_{II}/K_{IC} are smaller. Still the values of K_I/K_{IC} are smaller than those of

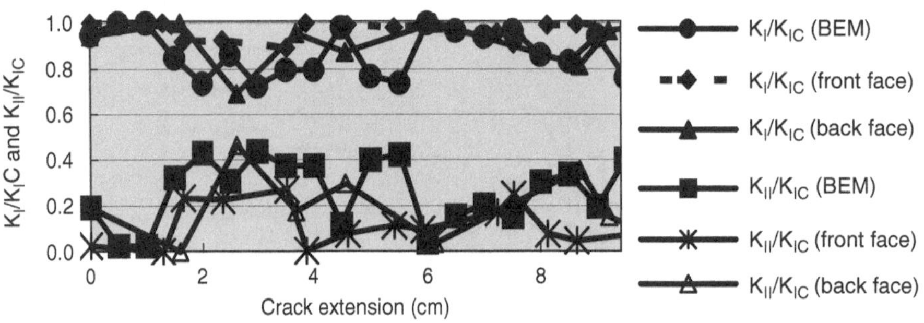

Figure 9: K_I/K_{IC} and K_{II}/K_{IC} during micro-cracking for diagonal crack D_1 in specimen-A.

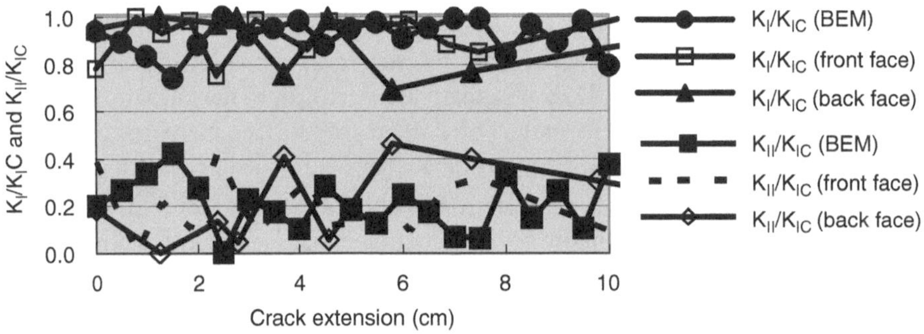

Figure 10: K_I/K_{IC} and K_{II}/K_{IC} during micro-cracking for diagonal crack D_2 in specimen-A.

the surface crack Sv. This implies that the mechanisms of the diagonal crack D_2 are in between those of cracks Sv and D_1.

6.4 Diagonal crack D_1 in specimen-B

Through all the cracking steps, crack trace due to uniform pressure is the closest to the actual trace in the experiment. Analyzed K_I/K_{IC} and K_{II}/K_{IC} are shown in Figure 11. In the beginning of the crack extension, K_I/K_{IC} is again nearly equal to 1.0. Sometimes K_{II}/K_{IC} slightly increases. It is concluded that for micro-cracking of the diagonal crack D_1, tensile cracks are dominant but mixed-mode cracks are sometimes active as confirmed by 3-D VRML visualization in Figure 7.

6.5 Digonal crack D_2 in specimen-B

Through all the cracking steps, crack trace due to uniform pressure is the closest to the actual trace in the experiment. Analyzed K_I/K_{IC} and K_{II}/K_{IC} are shown in

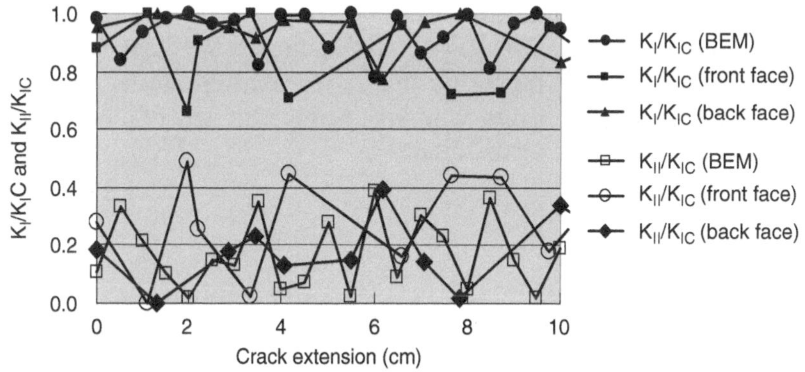

Figure 11: K_I/K_{IC} and K_{II}/K_{IC} during micro-cracking for diagonal crack D_1 in specimen-B.

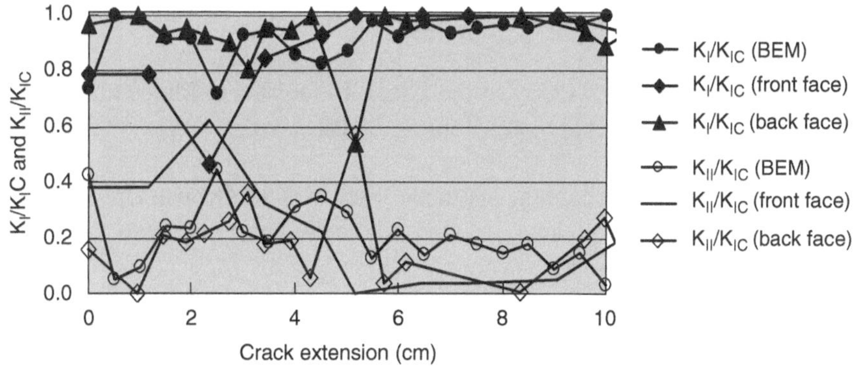

Figure 12: K_I/K_{IC} and K_{II}/K_{IC} during micro-cracking for diagonal crack D_2 in specimen-B.

Figure 12. In the beginning of the crack extension, K_I/K_{IC} is nearly equal to 0.8. K_{II}/K_{IC} becomes greater than K_I/K_{IC} in the beginning at the front face and in the middle at the back face of micro-cracking. It is concluded that for micro-cracking of diagonal crack D_2, tensile cracks are not only dominant but also shear and mixed-mode cracks are fairly active as confirmed by 3-D VRML visualization in Figure 7.

7 CONCLUSION

Micro-cracks due to rebar corrosion in concrete is studied analytically and experimentally. A two-domain BEM is applied to the mixed-mode crack extension based on the maximum circumferential stress criterion. Traces of the surface crack and the diagonal cracks in the arbitrary direction are analyzed. These cracks were produced by employing expansive agent into concrete specimen.

Micro-cracking mechanisms are investigated by BEM and confirmed by AE-SiGMA procedure with 3-D VRML. Depending on the crack types, contributions of mode I and mode II are varied during crack propagation. Micromechanics of corrosion cracking in concrete is mostly of mode I fracture along with mode II and the mixed-mode.

References

[1] Ohtsu, M., Simplified Moment Tensor Analysis and Unified Decomposition of AE source, Journal of Geophys. Res.; 96, pp.6211–6221, 1991.
[2] Ohtsu, M., Source Mechanism and Waveform Analysis of Acoustic Emission in Concrete, Journal of AE, 2(1), 103–112, 1982.
[3] Farid Uddin, A.K.M., and Ohtsu, M., BEM Analysis of Mixed-Mode Crack Propagation due to Corrosion of Reinforcement in Concrete, Journal of Materials, Concrete Structures and Pavements, JSCE, No.704/V-55, pp.271–280, 2002.
[4] Erdogan, F., and Sih, G.C., On the Crack Extension in Plates under Plane Loading and Transverse Shear, Journal of Basic Eng., No.12, pp.519–527, 1963.
[5] Carpinteri, A., Mechanical Damage and Crack Growth in Concrete, Martinus Nijhoff Publishers, Dordrecht, the Netherlands, 1986.

Earthquakes and Acoustic Emission – Carpinteri & Lacidogna (eds)
© 2007 Taylor & Francis Group, London, ISBN 978-0-415-44402-6

DAMAGE MECHANICS OF CARBONATED CONCRETE BY AE RATE-PROCESS ANALYSIS

T. Suzuki, G. Komeno & M. Ohtsu
Graduate School of Science and Technology, Kumamoto University,
Kumamoto 860-8555, Japan

ABSTRACT

Degradation of concrete structures could occur due to external effect as carbonation, which normally leads to changes physical and chemical properties of concrete. The degree of carbonation is, in most cases, evaluated as the depth of carbonation by spraying 1% phenolphthalein solution. For maintenance and management of concrete structures, it is desirable to evaluate not only the carbonation depth, but also the degree of damage. Quantitative damage evaluation of concrete is proposed by using acoustic emission (AE) method and damage mechanics. In the present study, carbonated concrete is examined, based on fracture behavior under uniaxial compression. Samples were prepared by accelerated carbonation process. AE behavior of concrete under compression is dependent on the degree of damage, and could be analyzed, applying the rate process theory. Using Loland's model, a relationship between stress and strain is modeled, and then a relation between AE rate and the damage parameter is correlated, constructing a database. By calculating the relative damage from the database, the damages of carbonated samples are quantitatively estimated.

1 INTRODUCTION

The durability of concrete structures varies depending on the external effect, as carbonation. The degree of damage in concrete is, in most cases, evaluated from mechanical properties. Among them, compressive strength has been frequently discussed as an index of degree of damage. Because, a concrete is a porous material, with the progress of carbonation, carbonated section increases accelerated. Micro-fractures occur due to stress concentration, though the average acting stress is small. Therefore, when discussing the damage of concrete under the carbonation process, it is rational to focus on the deterioration process of concrete. In the present study, chemical damage evaluation of carbonated concrete by acoustic emission (AE) measurement is discussed in a uniaxial compression test. Two kinds of specimens are prepared for evaluation; namely, concrete core samples subjected to accelerate carbonation with 10% CO_2 concentration and carbonated concrete core samples taken from an existing structure. Fracturing behavior of concrete is monitored by AE measurement. AE activity under uniaxial compression is analyzed by the rate process theory, and the damage parameter derived from the stress-strain behavior is evaluated by using Loland's model. These results are applied to estimation of Young's modulus of intact concrete and then the relative damage is calculated. The database is constructed to make the proposed procedure applicable to a limited number of samples taken from an existing structure.

2 ANALYTICAL PROCEDURE

2.1 AE rate process analysis

AE behavior of a concrete sample under uniaxial compression is associated with the generation of micro-cracks. These micro-cracks gradually are accumulated until final fracture that severely reduces load-bearing capacity. The number of AE events, which correspond to the generation of these cracks, increases acceleratedly. It appears that this process is dependent on the number of cracks at a certain stress level and the rate of the fracture growth, and thus is subjected to a stochastic process. Therefore, the rate process theory is introduced to quantify AE behavior under compression [1]. The following equation of the rate process is formulated to represent AE occurrence $\mathrm{d}N$ due to the increment of stress from V to $V + \mathrm{d}V$,

$$f(V)\mathrm{d}V = \frac{\mathrm{d}N}{N}, \tag{1}$$

where N is the total number of AE events and $f(V)$ is the probability function of AE at stress level V(%).

For $f(V)$ in Eq.1, the following hyperbolic function is assumed,

$$f(V) = \frac{a}{V} + b, \tag{2}$$

where a and b are empirical constants. Here, the value "a" is named the rate.

In Eq.1, the value of "a" reflects AE activity at a designated stress level, such that at low stress level the probability varies, depending on whether the rate "a" is positive or negative. In the case that the rate "a" is positive, the probability of AE activity is high at a low stress level, indicating that the structure is damaged. In the case of the negative rate, the probability is low at a low stress level, revealing that the structure is in stable condition. Therefore, it is possible to quantitatively evaluate the damage in a concrete structure by the rate process analysis of AE activity.

Based on Eqs.1 and 2, the relationship between total number of AE events N and stress level V is represented as the following equation,

$$N = CV^a \exp(bV). \tag{3}$$

Where C is the integration constant.

2.2 Loland's model

A damage parameter Ω in damage mechanics can be defined as a relative change in modulus of elasticity, as follows,

$$\Omega = 1 - \frac{E}{E^*}, \tag{4}$$

where E is the modulus of elasticity of concrete and E^* is the modulus of elasticity of concrete which is assumed to be intact and undamaged.

Loland assumed that the relationship between damage parameter Ω and strain ε under uniaxial compression is expressed [2],

$$\Omega = \Omega_{0+} A_0 \varepsilon^\lambda, \tag{5}$$

where Ω_0 is the initial damage at the onset of the compression test, and A_0 and λ are empirical constants.

The following equation is derived from Eqs. 4 and 5,

$$\sigma = (E_0 - E^* A_0 \varepsilon^\lambda)\varepsilon, \tag{6}$$

here,

$$E_0 = E^*(1 - \Omega_0). \tag{7}$$

2.3 Young's modulus E^* of the intact concrete using a database

As given in Eq.5, the initial damage Ω_0 in damage mechanics represents an index of damage. In Loland's model (Eq. 4), it is fundamental to know Young's modulus

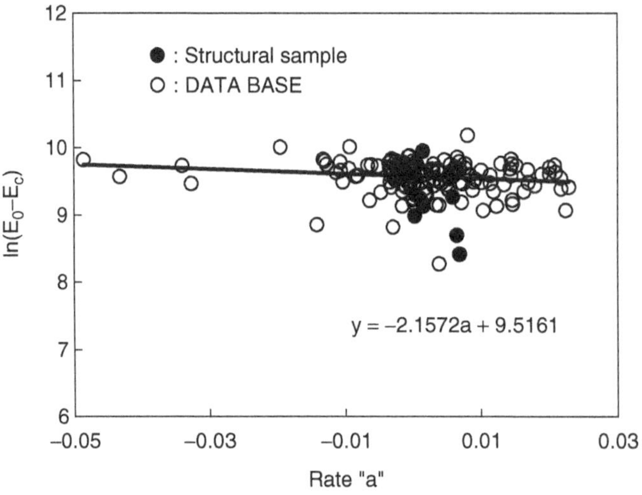

Figure 1: Relations between $\log_e(E_0 - E_c)$ and the ratio "a".

of the intact concrete (E^*). However, it is not easy to obtain E^* from an existing structure. Therefore, it is attempted to estimate E^* from AE measurement. One relation between total number of AE events and stress level and the other between stress and strain are taken into account. Based on a correlation between these two relations, a procedure is developed to evaluate the intact modulus. A correlation between the decrease of the Young's modulus under uniaxial compression, $\log_e(E_0 - E_c)$, and the rate "a" derived from AE rate process analysis is shown in Fig.1. These contain all the data, which were previously obtained in our laboratory as the database. Good correlation between $\log_e(E_0 - E_c)$ and the rate "a" value is confirmed [3]. From Eq.4, the decrease of Young's modulus under uniaxial compression ($E_0 - E_c$) is expressed, as follows,

$$E_0 - E_c = E^*(1 - \Omega_0) - E^*(1 - \Omega_c) = E^*(\Omega_c - \Omega_0) \tag{8}$$

Here Ω_c is the damage at the final stage. Based on a linear correlation in Fig.1,

$$\log_e(E_0 - E_c) = \log_e[E^*(\Omega_c - \Omega_0)] = Dx + C. \tag{9}$$

Here, it is assumed that $E_0 = E^*$ when $a = 0$. This allows us to estimate Young's modulus of intact concrete E^* from AE rate process analysis as,

$$E^* = E_c + e^c. \tag{10}$$

3 EXPERIMENTS

3.1 Specimens

Cylindrical samples were subjected to accelerate carbonation, and carbonated core samples were taken from an existing structure. Concrete samples (10 cm in diameter and 20 cm in height) were carbonated under 10% CO_2 concentrations. The accelerated carbonation was carried out according to JIS A 1153 (2003). Dimensions of core samples taken from an existing structure were 5 cm in diameter and 10 cm in height. The structure had been constructed in 1967. It is found that carbonation has progressed for these 36 years.

3.2 AE measurement

In uniaxial compression test of silicon grease was pasted on the top and the bottom of the samples, and a Teflon sheet was inserted to reduce AE events generated by friction. MISTRAS-AE system (PAC) was employed as a measuring device. AE hits were detected by using an AE sensor UT-1000 (resonance frequency: approx. 1 MHz). The frequency range was from 60 kHz to 1 MHz. To count the number of AE hits, the threshold level was set to 60 dB with a 40 dB gain in a pre-amplifier and 20 dB gain in a main amplifier. For event counting, the dead time was set as 2 msec. It should be noted that AE measurement was conducted with two channels as well as the measurement of axial and lateral strains.

4 RESULTS AND DISCUSSION

4.1 Mechanical properties of concrete due to carbonation

Carbonation develops from the surface of structural concrete. Due to carbonation, C-S-H and $Ca(OH)_2$ in concrete are transformed into $CaCO_3$. This could lead to the changes the mechanical properties and the pore distribution in concrete. It is found that the compressive strength increases with the progress of carbonation. The compressive strength increased by 40% after 42 days of CO_2 exposure. This is caused by the decrease in the pore ratio in mortar due to carbonation. Comparing pore diameters of carbonated concrete with that of normal concrete, it is observed that pore diameter with 0.03 μm over were dominant in carbonated concrete, while pore with 0.013 μm under were dominant in normal concrete. The pore ratio of carbonated concrete decreases with the decrease in the number of micro-pores. Carbonation develops as CO_2 dissolves into H_2O and changes into HCO^{3-}. Thus, as micro-pores hold moisture in a porous space, pore closure could occur.

Due to carbonation, the stiffness decreases while the strength increases. This phenomenon occurs because the number of pores with small diameter decreases due to carbonation, changing the distribution of pore diameters. Thus, the progress of carbonation not only causes a decrease of pH, but also changes material properties. It is considered that micro fractures occur under these circumstances due to stress concentration, though the average stress is small.

4.2 AE generating behavior in carbonated concrete

AE generating behavior in carbonated concrete was formulated using the rate process analysis. One relation between the probability function of AE activity and the stress level is shown in Fig.2. The rate process analysis was conducted at stress level in the range from 30% to 80%. This is because AE events occurring at initial loading below 30% strength due to contact with the loading plate and at an accelerated stage above 80% has little to do with the damage.

AE activity of carbonated concrete was higher than that of normal concrete (after 28-day curing) at each stress level. It is considered that in carbonated concrete under compressive stress, micro cracks are generated with an increase in strain, thus increasing the number of AE events. Based on results of AE rate process analysis, it is concluded that carbonation of concrete is leads to the changes in the pore structure caused by the transform of cement hydrate materials.

4.3 Relative damage evaluation

Quantitative damage evaluation of carbonated concrete was carried out, based on the estimated Young's modulus. It is demonstrated that Young's modulus of intact concrete E^* can be estimated by Eq.10 [4]. Thus, to estimate relative damage of the concrete, ratios E_0/E^* are obtained. To determine E^* from the relation in Fig.1, it is necessary to analyze a large number of specimens. However, the

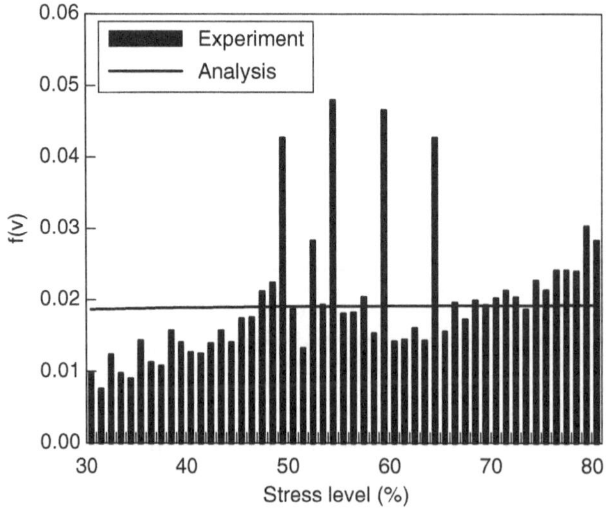

Figure 2: Relations between $f(V)$ and stress level.

number of concrete cores available is limited in existing structures. Therefore a database, which could allow even a single concrete core to be evaluated, is constructed as shown in Fig.1. Relative damage evaluation of carbonated concrete is shown in Fig.3. With the passage of exposure periods, the evaluated values decrease. From these results, it is considered that the degree of damage advances in carbonated concrete, as strain increases under compressive stress. Relations between the carbonation depth and the relative damage are summarized in Fig.4. When the carbonation depth is less than 5 mm, the values obtained

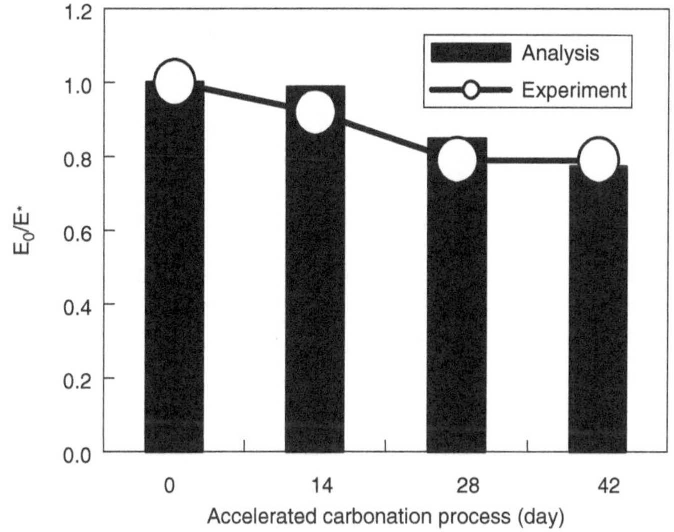

Figure 3: Relative damage evaluation.

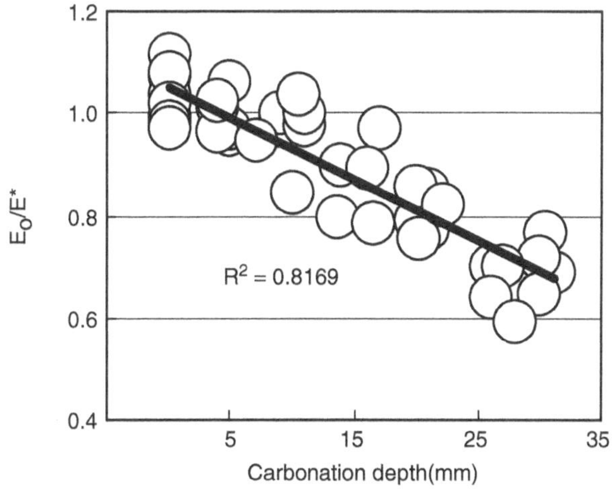

Figure 4: Relations between carbonation depth and E_0/E^*.

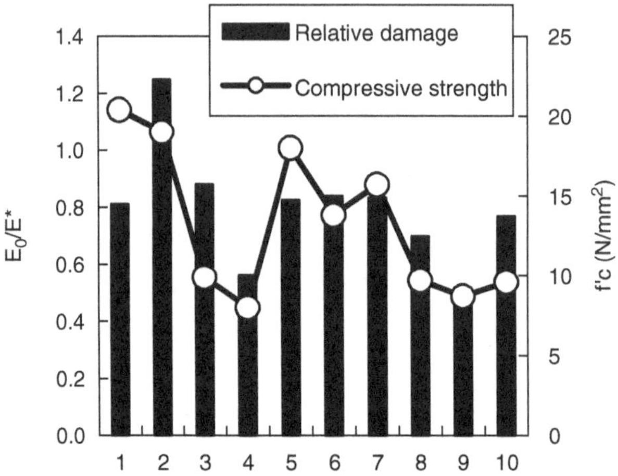

Figure 5: Relations between E_0/E^* and compressive strength in structural concrete.

from the relative damage evaluation exceed 1.0, implying no damage. As the carbonation depth increases, the relative damage evaluated decreases reasonably. This means that the deterioration due to chemical effect as carbonation can still be estimated by AE rate process analysis, although the strength increases due to carbonation. Relations between the relative damage and the compressive strength of carbonated concrete cores taken from an existing structure are shown Fig. 5. Relative damage, E_0/E^*, of concrete is obtained as equal to 1.0. Relative moduli E_0/E^* vary from 0.48 to 1.25, as those 9 samples are estimated below 1.0. Though the degree of damage in the core samples is not clearly identified by the compression strength, the deterioration is quantitatively estimated, using the relative damage evaluated by AE measurement.

5 CONCLUSION

Quantitative damage evaluation of carbonated concrete is studied by AE rate process analysis. AE generating behavior in carbonated concrete is closely related with the pore structure of concrete, which is analyzed based on the rate process theory and damage mechanics. Thus, the degrees of damage in concrete samples are quantitatively evaluated. Conclusions are summarized below.

(1) AE generating behavior in carbonated concrete is analyzed by applying the rate process theory.
(2) The relative damage with the progress of carbonation. This could result from an increase in the number of micro-cracks, as strain increases due to carbonation.

(3) In concrete cores damaged by carbonation, the relative damage evaluated and the compressive strength show a comparable trend.
(4) The deterioration due to chemical effect as carbonation can still be estimated by AE rate process analysis.

References

[1] A. Ishibashi, K. Matsuyama and M. Ohtsu (1998), "AE Application for Diagnosis of Deteriorated Concrete of Harbor Structures", Proc.6th Int.Sym.on AE from Composite Materials, 145–152.
[2] Loland, K.E. (1989), "Continuous Damage Model for Load – Response Estimation of Concrete", Cement and Concrete Research, Vol.10, 395–402.
[3] T. Iida, H. Watanabe, Y. Tomoda and M. Ohtsu (2000), "Damage Estimation of Concrete Core by AE Rate Process Analysis", Proc. of the Japan Concrete Institute, Vol.22 (1), 271–276.
[4] T. Suzuki, H. Watanabe and M. Ohtsu (2002), "Damage Evaluation in concrete Using Acoustic Emission Method", The 6th Far-East Conference on Non-Destructive Testing, 111–116.

Earthquakes and Acoustic Emission – Carpinteri & Lacidogna (eds)
© 2007 Taylor & Francis Group, London, ISBN 978-0-415-44402-6

BRAZILIAN TEST OF CONCRETE EVALUATED BY AE

H.W. Reinhardt, F. Finck, C. Grosse & J. Kurz
University of Stuttgart, Institute of Construction Materials, Pfaffenwaldring, 70569 Stuttgart, Germany

ABSTRACT

For a detailed evaluation of fracture processes due to tensile stresses using acoustic emission (AE) technique, a splitting test (Brazilian test) is particularly well-suited. Thus, a Brazilian test was performed to generate a well-defined crack surface in a concrete cube. Using AE technique, the formation of micro cracks was observed and correlated with the stress distribution during the test. During the experiment, some 1800 acoustic emissions were recorded. Half of these events accompanied the evolution of the main crack. From these 900 events, 378 AE events were localized with an accuracy better than 5 mm.

The well-defined crack in the center of the specimen allowed a mapping of the crack surface. For this purpose, the specimen was ground in a stepwise manner normal to the crack subsequent to testing. The run of the crack in the ground surfaces was digitized from photographs and interpolated. Inversion of AE amplitudes of the spatial displacement distribution yields a system of equivalent moments for selected AE events, the so-called moment tensors. Analysis of the moment tensor allows an interpretation of failure based on fracture mechanics and gives information about the stress field within the specimen.

1 INTRODUCTION

For a detailed evaluation of fracture processes due to tensile stresses using acoustic emission technique, a splitting test (Brazilian test) is particularly well-suited (Finck et al. [1]). During the experiment presented in this article some 1800 acoustic emissions were recorded. Half of these events accompanied the evolution of the main crack. From these 900 events 378 were localized with an accuracy better than 5 mm. The well-defined crack in the center of the specimen allowed a mapping of the crack surface. For this purpose, the specimen was ground in a stepwise manner normal to the crack subsequent to testing. The run of the crack in the ground surfaces was digitised from photographs and interpolated. An advanced 3D visualization routine combines the results of the different analysis methods.

2 EXPERIMENTAL SETUP

To generate controlled tensile failure, a splitting test (DIN 1048 [2]) with a concrete cube (edge length 200 mm) was performed. Compressive load was applied using two parallel steel edges, one on the top and the other on the bottom of the specimen. This produced a tensile crack parallel to the steel edges. A controlled opening of the tensile crack is necessary to obtain as many acoustic emissions during the evolution of the crack as possible. Therefore, the crack opening was measured by two linear displacement sensors (LVDT), one on each side of the cube. The mean crack opening acted as the control parameter for the loading device. Acoustic emissions were registered by eight piezo-electric ultrasonic sensors and recorded by a transient recorder with a sample frequency of 5 MHz per channel and an amplitude resolution of 12 Bit. Additionally, piston displacement, load and crack opening were recorded. According to DIN EN 206, the concrete had a strength grade of C35/45. The maximum aggregate size was 16 mm. Before the test, a compressive wave (P-wave) velocity of 4600 m/s was determined for the specimen.

3 EVALUATION OF RESULTS

The analysis of the data obtained during the splitting test can be divided into four categories:

(1) Analysis of mechanical data and the acoustic emission rate,
(2) Localization of acoustic emissions,
(3) Evaluation of the topography of the fracture plane,
(4) Fracture mechanical studies on moment tensors.

In the following sections, results of the experiment are presented and discussed.

3.1 Mechanical test data

In Figure 1 the mechanical test data and the acoustic emission rate over the test period are illustrated. The first plot contains the load as a function of the mean crack opening. It can be seen that the first phase shows a more or less linear elastic behavior until the load decreases with an increasing crack opening. Towards the end of the test the crack opens with a behavior similar to creep at a nearly constant load.

The second and third plots show the load over time and displacement over time behavior, respectively. LVDT 2 was attached on the backside of the specimen, LVDT 1 on the front. The crack opening started from the back of the specimen and grew through the cube to the front side. With an advanced opening of the

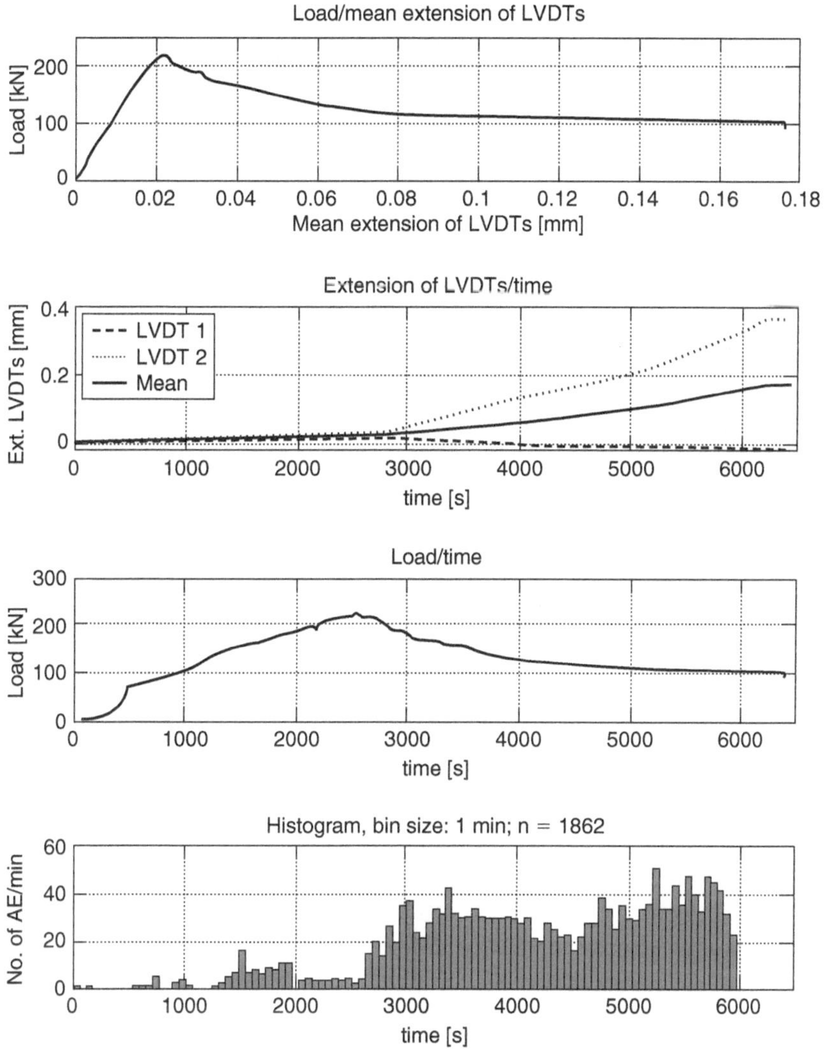

Figure 1: Mechanical test data and a histogram of the acoustic emission activity during the test.

crack, LVDT 1 showed negative values. The negative values occurred because the deformation of the specimen led to a compression of LVDT 1. No crack was visible on the front side of the specimen until the end of the experiment. In the load versus time curve, the compression of the felt strip between the specimen and the steel edges at the application of load can be seen during the first 500 seconds. After that time, load increased nearly linearly, with a small break at about 1500 seconds, which is also accompanied by acoustic emissions (see histogram).

The histogram on the bottom of Figure 1 gives the acoustic emission rate per minute. The trigger times of all registered acoustic emissions are taken into account. Because the rate of events was well-below the capacity of the acquisition system, it can be assumed that no data was missed. In addition to the mentioned activity reflected in the minor break in the ascending branch of the load versus time curve, the acoustic emission rate increases after the achievement of maximum load. From that point the crack grows through the specimen, accompanied by numerous acoustic emissions. The splitting tensile strength amounted to 3.52 MPa.

3.2 Localization

Acoustic emissions are observed by a number of sensors with well-defined positions and the Pwave velocity has to be measured. For the estimation of the focal coordinates of an acoustic emission the accurate onset times of the P-waves must be determined. A simple system of linear equations is the basis for this analysis. Since the system of equations is overdetermined when data from eight sensors is available, the run time residuals are minimized with the method of least squares. This leads to improved accuracy and yields error bounds for the estimated focal coordinates (Grosse et al. [3]). From the 1800 acoustic emissions registered, the onset times of the first 900 events were picked with the WinPecker© software (Grosse, [4]). From these events, 378 could be localized with an accuracy better than 5 mm.

In Figure 2, results of the localization are illustrated. The events are projected onto the main coordinate planes revealing a view from above, from the front and the right side of the specimen. The brightness of the data points provides information about the time when the events occurred. Light grey represents events from the beginning of the test, dark grey represents events from an advanced stage. The last localized event occurred at 4246 seconds after the start of loading. The acoustic emissions are located along a well-defined zone of failure parallel to the steel edges applying compressive load. Local damage at the point

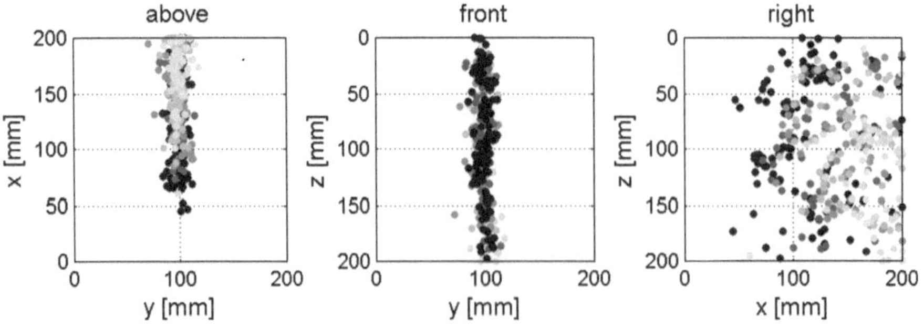

Figure 2: Projections of the localization of acoustic emissions onto the x/y, y/z and x/z planes. Events marked in light grey occurred at the beginning of the test, dark ones at an advanced stage of loading.

of load application was minimized through the use of felt strips which evenly spread the applied compressive load. Consequently, only a small number of acoustic emissions were visible in the region where load was applied. It is possible, however, that some of the acoustic emissions from this regions were ignored by the localization process due to a weak signal to noise ratio. Considering the brightness of the data points, the growth of the crack from the back to the front is clearly visible. Even at an advanced stage of the test, where the crack front had progressed through the specimen, acoustic emissions occurred at the back side of the specimen, where the crack was open already. This can be explained by either an increasing accretion of micro cracks or a rubbing of the crack surfaces.

3.3 Topography of the fracture plane

The splitting test performed revealed a well-localized and visible crack in the middle of the specimen. For a correlation of this macroscopic crack with the localization of the acoustic emissions (i.e. micro cracks) along that zone of failure, the topography of the crack surface was investigated. The specimen was ground from the top in steps of 5 mm. The surfaces of the various steps were photographed. From the digitized run of the crack in every section, a fracture plane was interpolated. The digitization process of the crack has an error of approximately 2 mm. In most of the sections only a single crack was visible. The crack went through the concrete matrix from one aggregate to the next. Depending on the strength of the grains, they were divided or the crack ran around them, which caused the crack surface to undulate. The localization of the acoustic emissions revealed, that micro cracking and debonding also occurred in a distance of several millimeters to the final macro crack.

3.4 Fracture mechanics

The basics of evaluating the moment tensor have been presented in the references (Andersen [5], Finck [6], Grosse [7], Jost and Hermann [8]). Two clusters containing 5 and 9 events, respectively, were selected from the localized events. Since the Green's functions, representing the transfer functions of the medium are eliminated in the relative moment tensor inversion, the radii of these clusters have to be small enough and the events should originate in a relatively short temporal interval. Low frequency noise from the data was minimized by a suitable wavelet filter algorithm. The transient raw data, representing velocity over time, were then integrated to reveal the displacement. The first half-wave amplitudes of the longitudinal P-waves were picked from the data for input.

An overview of the results of the moment tensor inversion is presented in the form of the radiation patterns in stereographic projections (Figure 3). All events in cluster C2 show a similar behavior, with tension (dark) more or less parallel

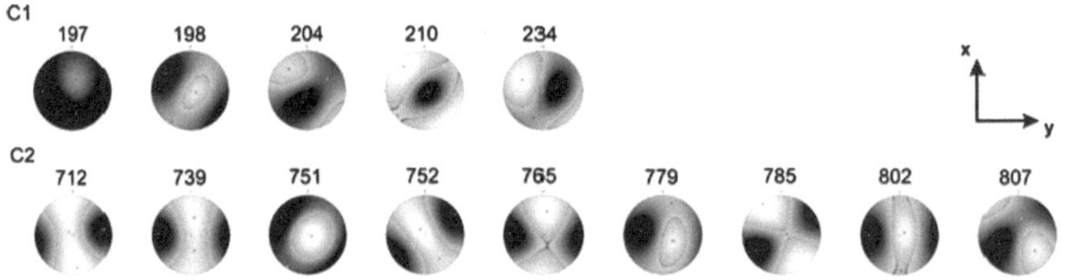

Figure 3: Radiation patterns of all moment tensors from clusters C1 and C2.

to the y-axis. A small compressive component almost parallel to the z-axis is indicated by negative amplitudes (light colored).

The decomposition of the moment tensors is revealing the percentile contributions of isotropic, double couple and CLVD mechanisms. Significant positive isotropic components and a mixture of double couple and CLVD mechanisms are evident. The solutions for cluster C1 are not that consistent. The directions of positive amplitudes due to tensile stress vary more around the y-axis than in C2. The decomposition shows similar results as in cluster C1. Event 210 has a negative isotropic component, which could be explained by a pore collapse in combination with shear.

The results corroborate the assumption of mode I failure with a major tensile crack parallel to the steel edges from which load was applied. The mean tensile stress axis is parallel to the y-axis, which is also the preferred direction of positive amplitudes.

4 SUMMARY

A splitting test was performed to generate controlled failure in a concrete specimen. During the test, acoustic emissions were recorded for a detailed investigation of the damage zone. The configuration of the stress field produced a well-defined crack, which allowed for the mapping and digitization of the topography of the crack surface. The macro crack ran from the back to the front side of the specimen. This was observed in the mechanical data, as well as in the results of the localization.

For an evaluation of the stress field and the fracture mechanisms leading to failure, moment tensors were calculated. In the moment tensors a significant positive isotropic component was observed and the deviatoric component consisted primarily of a double couple. The direction of maximum tension in the radiation patterns was almost parallel to the mean tensile stress generated by the experimental setup. A small compressive component was found parallel to the z-axis.

The investigations reveal tensile fracturing as the dominant cause of failure. However, variations of the tension axes and deviations from pure tensile failure are observed. Inhomogeneities, such as aggregates, generate undulations of the crack surface and can be responsible for an enhanced complexity of micro cracking. Corresponding shear stresses would explain the significant double couple components.

The projection of the seismic radiation patterns onto focal spheres in a 3-dimensional image depicts the results of the moment tensor inversion. These projections can be combined with the results of the localization and other investigations and allows for a comprehensive understanding of failure mechanisms in a specimen.

Signal-based analysis of acoustic emissions is of course more time consuming than the classic parameter-based approach, where a range of parameters is extracted automatically and a rough localization is sometimes possible. However, signal-based analysis allows for much more detailed evaluation of failure processes. Thus, the accuracy of the localization is enhanced. Furthermore, the localization can be automated for a number of events using suitable de-noising algorithms and automatic onset pickers. Important knowledge about the stress field and fracture mechanics can be obtained from the data. A broad understanding of the radiation pattern of seismic energy and an automation of the data processing is the basis for future applications in structural health monitoring.

ACKNOWLEDGEMENT

This work was funded by the Deutsche Forschungsgemeinschaft DFG (SFB 381, TP A6).

References

[1] Finck, F., Yamanouchi, M., Reinhardt, H.-W. and Grosse, C.U.: Evaluation of mode I failure of concrete in a splitting test using acoustic emission technique. Int. Journal of Fracture, 2004 (in print).

[2] DIN 1048, part 5: Prüfverfahren für Beton – Festbeton, gesondert hergestellte Probekörper. Deutsches Institut für Normung e. V, 1991.

[3] Grosse, C.U., Reinhardt, H.-W. and Finck, F.: Signal-Based Acoustic Emission Techniques in Civil Engineering. Journal of Materials in Civil Engineering (ASCE) 15 (3), pp. 274–279, 2003.

[4] Grosse, C.U.: WinPecker – Programm zur vollautomatischen dreidimensionalen Lokalisierung von Schallemissionsquellen. In: 12. Kolloquium Schallemission, Jena. DGZfP Berichtsband 72, pp. 191–204, 2000.

[5] Andersen, L.M.: A relative moment tensor inversion technique applied to seismicity induced by mining. Ph.D. thesis, University of the Witwatersrand, Johannesburg, South Africa, 2001.

[6] Finck, F.: Application of the moment tensor inversion in material testing. Otto-Graf-Journal 12, pp. 145–156, 2001.

[7] Grosse, C.U.: Grundlagen der Inversion des Momententensors zur Analyse von Schallemissionsquellen. In: Werkstoffe und Werkstoffprüfung im Bauwesen. Festschrift zum 60. Geburtstag von Prof. Dr.-Ing. H.-W. Reinhardt. Libri BOD, Hamburg, pp. 82–105, 1999.

[8] Jost, M.L. and Hermann, R.B.: A students guide to and review of moment tensors. Seism. Res. Letters 60, pp. 37–57, 1989.

2.3 AE testing on steel, composites and other materials

Earthquakes and Acoustic Emission – Carpinteri & Lacidogna (eds)
© 2007 Taylor & Francis Group, London, ISBN 978-0-415-44402-6

EVOLUTION OF DAMAGE ACCUMULATION IN LOW-CARBON STEEL IN TENSION CONDITION

M.R. Tyutin[1], L.R. Botvina[1], N.A. Zharkova[1], T.B. Petersen[2] & J.A. Hudson[3]
[1]A.A. Baikov Inst. of Metallurgy and Materials Science, Moscow, Russia
[2]PRC Kurchatov Institute, Moscow, Russia
[3]Cambridge University, UK

ABSTRACT

Stages of damage accumulation process in notched specimens from low-carbon steel at tension have been studied by means of replicas, acoustic emission and ultrasound attenuation methods. The shape and size of plastic deformation zones at a pre-existing crack tip, microcrack length and density in the zones have been estimated and cumulative distributions of microcracks by their size have been plotted. Using acoustic emission method, number and amplitude of acoustic emission signals characterizing damage accumulation process on various stages of loading and also location of the signals on the specimen surface have been estimated and cumulative amplitude distributions of the signals have been plotted. It has been established that cumulative distributions of microcracks and acoustic emission signals on initial stage of damage accumulation are well described by an exponential function which is replaced by a power law one at a load close to the maximum value. At this stage of loading, density of microcracks and acoustic emission signals approaches the limited value, crack coalescence leading to the main crack formation begins and ultrasound attenuation coefficient sharply increases. The results obtained are compared to the regularities of fault accumulation in the earth crust.

1 INTRODUCTION

The process of fracture is multistage and involves the stages of nucleation, accumulation and growth of defects; their interaction; formation of the main crack; and the final fracture. Among the stages listed, the last stage of localized fracture described in the context of the fracture mechanics is the most extensively studied. The stage of multiple or distributed fracture, when the accumulation and growth of different-scale defects occurs, is the least understood because it typically requires the use of sophisticated techniques for the evaluation of defects of various kinds. However, this may be labour well spent, since an analysis of the damage

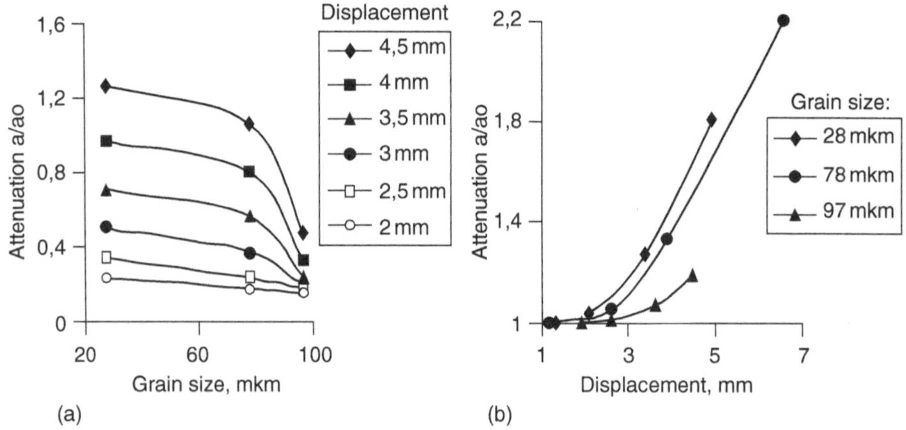

Figure 1: Dependence of a relative ultrasound attenuation coefficient on a grain size and deformation of notched specimen from low-carbon steel.

accumulation process in the materials belonging to different classes and scales has many common features allowing considering this process in terms of the theory of critical events. Therefore the main goal of this paper is establishing common peculiarities preceding critical event (main crack formation or fracture) observed by various methods and characterizing mechanical and physical behavior of material.

2 ESTIMATION OF ULTRASOUND ATTENUATION

Ultrasound attenuation coefficients were estimated at various stages of loading and after loading (Botvina et al. [1]). Measurements have been carried out in the plastic deformation zone near a crack tip at frequency of 5 MHz or under specimen fracture surface at range of frequencies from 2 to 20 MHz. The ultrasound attenuation coefficient increased with displacement and decreased with decreasing grain size of specimen material (Fig. 1). In Fig. 2, a dependence of attenuation coefficient at various stages of loading on microcrack density is presented. The microcrack density has been evaluated using relation (1) from studies of (Peacock et al. [2, 3]):

$$\mathrm{CD} = \frac{\mathrm{N} \cdot \mathrm{l}^3}{\mathrm{S} \cdot \mathrm{t}}, \tag{1}$$

where N is the number of cracks, l is the mean length of cracks, S is the surface area, t is the specimen thickness.

As can be seen from Fig. 2, the attenuation coefficient is linearly related to the microcrack density estimated at different displacements corresponding to various loading stages. Similar relationship between the attenuation coefficient

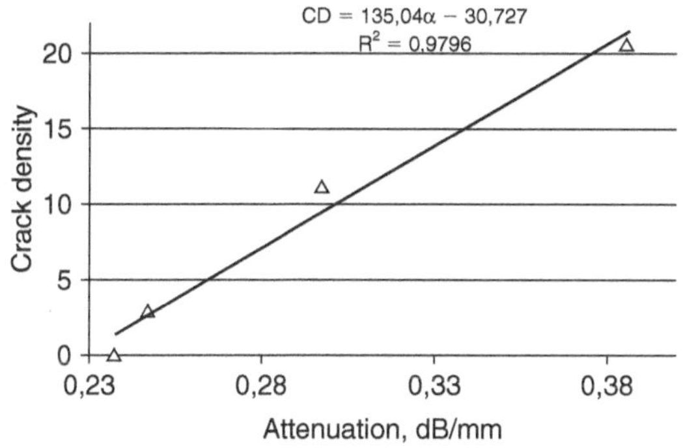

Figure 2: Dependence of microcrack density at various stages of tension on ultrasound attenuation.

and a crack density in specimens from marble has been obtained in [Peacock et al. [2]) where the accordance of experimental data on ultrasound attenuation with Hudson's theory (Hudson [4]) has been verified (see also Piau [5] and Peacock & Hudson [6]). Further verification of this relationship has been provided by Saenger et al. [7], using three-dimensional numerical studies.

3 ACOUSTIC EMISSION DATA

In Fig. 3 (a, b, c), the maps of acoustic emission signals location at various loading stages together with the changes of plastic zones that caused the acoustic activity are presented (Botvina et al. [1]). It can be seen that acoustic emission data correspond well to the increase in plastic zone which contain branches of localized deformation on initial stages (Fig. 3, d, e). With further increase in load, these branches merge and the second highly deformed plastic zone appears near crack tip of the newly formed main crack (Fig. 3, f).

A good correlation between the acoustic emission data and microcrack density within plastic deformation zones was revealed (Fig. 3). From Fig. 3 it follows that both parameters, the microcrack density and the corresponding number of acoustic emission signals, increase and reach limiting values with increasing displacement. The higher limited values of the characteristics correspond to the lower specimen thickness (Fig. 4, b). It means probably that the rate of microcrack initiation decreases as a result of their coalescence near the maximum tensile load, and such a decrease depends on the local stress state at the tip of a crack forming in the notched specimen near the maximum load. Acoustic emission captures this process well. Similar dependence of a crack density on a displacement is used in the model describing crack population in brittle rock materials (Fig. 4, c) (Scholz [8]).

The coalescence of cracks leads to a change in the function describing cumulative distributions of microcracks by their size and acoustic emission signals by their

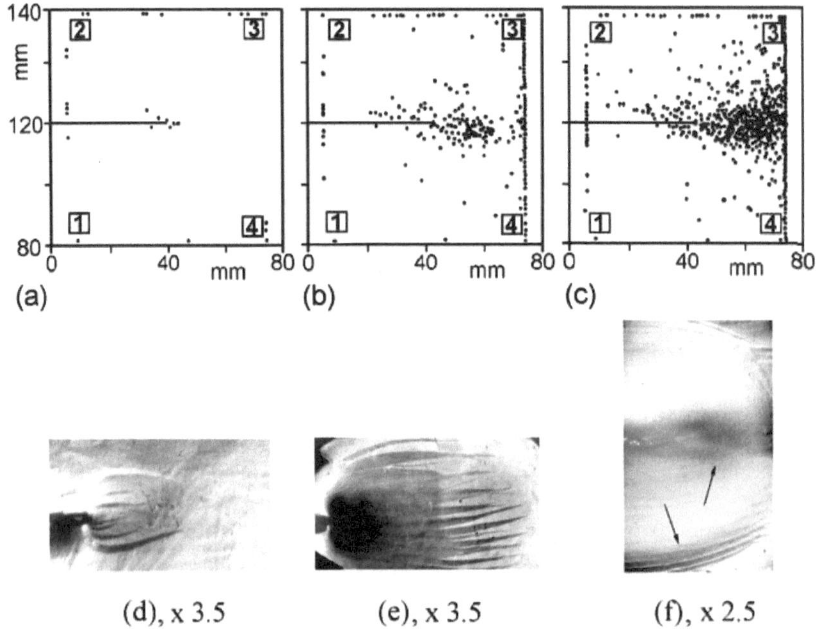

Figure 3: Results of acoustic emission signal location at various stages of tension of notched specimens (a, b, c) and corresponding plastic zones obtained by means of replicas method (d, e, f).

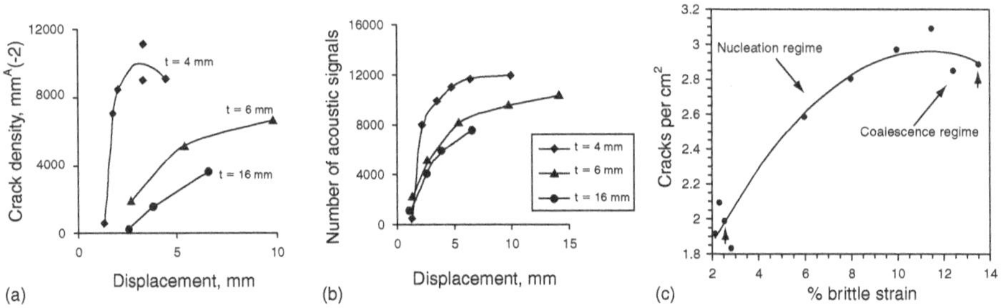

Figure 4: Dependences of microcrack density (a) and number of acoustic emission signals (b) on deformation [1] and development of population of faults in model of Spyropoulos (Scholz [8]).

amplitude (Fig. 5). It was found that both type of curves, different a nature, manifest the same changes with load increasing. At initial stages of loading, the curves are best fitted by an exponential function that is replaced by the power law one near the maximum load of fracture. This transition occurs at lower displacement and loads when specimens of larger thickness are used. At an intermediate stage of loading, the cumulative curves may be described both by an exponential function or a power law function with the same correlation coefficient.

Thus, the shape of the cumulative distributions of microcracks and acoustic emission signals is defined, probably, by the local stress state at a crack tip in a

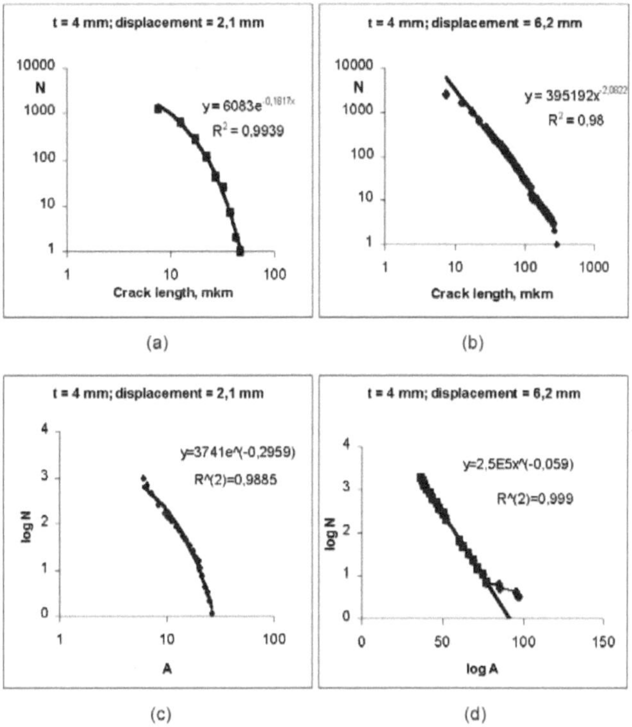

Figure 5: Cumulative distributions of microcracks (a, b) and acoustic emission signals (c, d) on an initial stage of plastic deformation (a, c) and near the maximum load (b, c).

specimen and a "distance" to a critical event of the process, namely, to the formation of the main crack at the maximum fracture load. As a "distance" to a critical event one may use the difference ($\varepsilon_c - \varepsilon$) of the critical deformation at the maximum load (ε_c) and the current deformation (ε).

Considering the fracture process in terms of the phase transition theory, this difference may serve as the order parameter of a phase transition during fracture. It has been established that a decrease in this parameter leads to the reduction in the exponent of the cumulative distribution curves characterising the damage accumulation process before the main crack formation. Similar change of the types of cumulative distributions has been found during studying the natural faults in the oceanic or earth crust. So, the authors of (Cowie et al. [9]) showed that cumulative fault distribution in the oceanic crust is well fitted by an exponential function, but this function becomes the power law one in the zone of active faults in the earth crust. It shows that the regularities of fault development in the earth crust, i.e. on a global scale level, are probably similar to those observed at studying microcrack accumulation in metallic specimens.

The appearance of power law dependence upon reaching of a critical state agrees with the theory of phase transitions according to which the power law function describes phenomena and processes occurring near the critical event.

However, many researches report the other sense of change in the functional dependence of the cumulative crack (or fault) distributions. So, in (Scholz [8]) it has been reported that the power law function describes cumulative distribution at the initial stage of fracture development, when the process of initiation and development of individual faults dominates. With increasing fault density, the fault interaction begins, their number decreases, and cumulative distribution becomes a power law one. These differences in the order of change in the functions describing cumulative crack distributions observed in specimens and natural faults may be connected to the difficulties of determining the stage of maximum damage accumulation under natural conditions.

Further investigations must bring to light the order of change in the functions describing the cumulative crack distributions plotted on various scales and corresponding to materials with different structure and strength.

ACKNOWLEDGEMENTS

This work was partially supported by the INTAS (project N 01-5-748) and by the Russian Foundation for Basic Research (project N 02-05-65231-a).

References

[1] Botvina L.R., Zharkova N.A., Tyutin M.R., Petersen T.B. and Budueva V.G., Diagnostics of damage accumulation in low carbon steel, Metals, 2004, (in press).

[2] Peacock S., McCann C., Sothcott J. and Astin T.R., Experimental measurements of seismic attenuation in microfractured sedimentary rock. Geophysics, 59, No. 9, 1342–1351, 1994.

[3] Peacock S., McCann C., Sothcott J. and Astin T.R., Seismic velocities in fractured rocks: an experimental verification of Hudson's theory. Geophysical Prosp., 42, 27–80, 1994.

[4] Hudson J.A., Overall properties of a cracked solid, Math. Proc. of the Cambridge Phil. Soc., 88, 371–384, 1980.

[5] Piau, M., Attenuation of a plane compressional wave by a random distribution of thin circular cracks. Int. J. Eng. Sci. 17, 151–167, 1979.

[6] Peacock, S. and Hudson, J.A. Seismic properties of rocks with distributions of small cracks. Geophys. J. Int. 102, 471–484, 1990.

[7] Saenger E.H., Kruger O.S., Shapiro S.A. Effective elastic properties of randomly fractured soils: 3D numerical experiments. Geophys. Prosp. **52**, 183–195, 2004.

[8] Scholz C.H., Earthquakes and Faulting, Cambridge Univ. Press, 2002.

[9] Cowie P.A., Malinverno A., Ryan W.B.F., Edwards M.N., Quantitative fault studies on the East Pacific Rise: a comparison of sonar imaging techniques. J. of Geoph. Res., **99** 15205–15218, 1994.

Earthquakes and Acoustic Emission – Carpinteri & Lacidogna (eds)
 ISBN 978-0-415-44402-6

FATIGUE PERFORMANCE OF SANDWICH COMPOSITES

B. Shafiq, A. Quispitupa & F. Just
School of Engineering, University of Puerto Rico at Mayagüez, USA

ABSTRACT

This paper delineates the results of fatigue testing performed to characterize, i. tensile fatigue lifetime and ii. flexure fatigue lifetime as a function of the thickness of foam core. Sandwich composites were made of single ply carbon fiber facesheets with urehane foam core. Acoustic emission analysis indicated core damage to be the predominant failure activity while fiber rupture served as a precursor to catastrophic failure. Near crack tip region was observed to have multiple crack initiation sites before the onset of crack growth. Crack path was found to erratic on the facesheets while the crack in the foam primarily followed a planar growth along the interface with the facesheets. Both mode I and II cracking was observed in the core and along the interface between the core and the facesheets. Flexural fatigue life was found to be unlimited below 75% of the ultimate static load. Foam core size effect is found to be significant in the crack growth and lifetime behavior. An AE based stiffness reduction model was developed to quantify the extent of damage and stochastic analysis was performed to account for large scatter in the lifetime data.

1 INTRODUCTION

Fatigue damage in sandwich composites is generally difficult to observe until the onset of catastrophic failure [1]. The detection, however, is greatly facilitated with the use of acoustic emission (AE) technique as it permits continuous damage inspection, classification and identification of modes of failure in various constituents of the composite in real time [2,3]. However, in spite of extensive use of AE technique in engineering applications [2–6]; literature on its application in fatigue crack growth (FCG) and especially in sandwich composites is scarce [1].

Sandwich composites are generally designed to carry flexural loads, however, accidental impacts, voids and micro-cracks inherent in sandwich composites can inevitably subject the component to tensile loading conditions. Therefore, an effort related to tensile fatigue characterization is outlined in this paper. In addition,

Figure 1: Test Setup.

preliminary results of an ongoing flexural fatigue process as a function of foam thickness are also presented.

2 RESULTS

Sandwich composite beams made of 1-ply facesheet of 0.5 mm thickness 161 g plain weave epoxy matrix carbon fiber and a core material of 6.0 mm thick urethane foam filled kraft paper honeycomb bonded to the facesheets with 1.5 oz vinyl ester resin were used in this study. Tensile fatigue testing was performed on SEN specimens of dimensions 250 mm × 38 mm × 12.7 mm (60° notch depth) under three point bending. Flexural fatigue testing was performed on specimen of similar dimensions but without a notch and thickness varying from 3 mm to 13 mm.

Flexural fatigue tests were performed between stress levels of 60 and 95% of the ultimate static load at a load ratio of 0.1 and a frequency of 2 hz. Fatigue life was observed to be unlimited below 75% of the ultimate static strength. The test setup of a specimen under tensile loading conditions is shown in Fig. 1.

With the analysis of AE events, energy and position, damage was classified in various constituents of the sandwich composite; such as, core, interface between core and the facesheets, resin and facesheets. Fig. 2a shows the load-deflection curve, while Figs. 2b shows corresponding energy vs amplitude curves for a typical specimen tested under quasi-static loading conditions.

Based on AE analysis under quasi-static and fatigue testing, damage was classified in various constituents of the sandwich composite as a function of AE parameters as presented in Table 1. Amplitude and energy level were found to be solely a function of material composition and independent of specimen geometry or mode

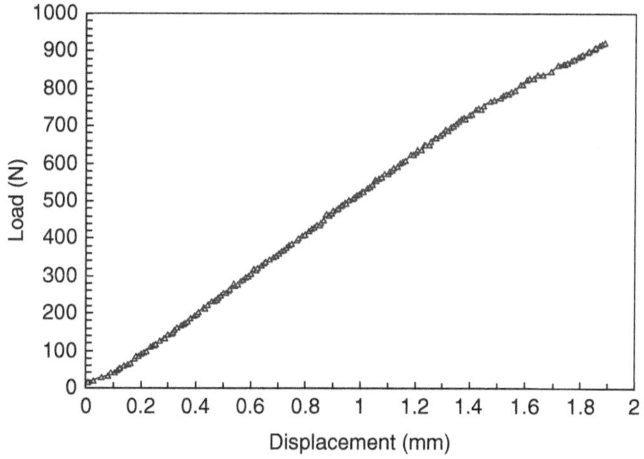

Figure 2a: Typical Load vs Displacement Curve.

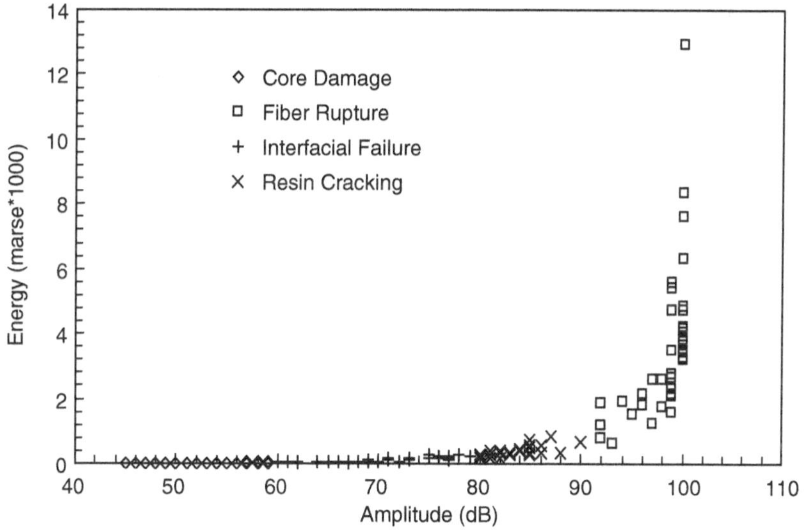

Figure 2b: Energy vs Amplitude corresponding to Fig. 2a.

Table 1: Sequence of failure and corresponding amplitude and energy ranges.

Failure Mode	AE Amplitude (dB)	AE Energy (marse)
Core Damage	45–59	0–25
Interface Failure	60–79	3–219
Resin Cracking	80–89	88–374
Fiber Rupture	Above 90	347–13568

of failure. This classification matched reasonably with the qualitative results reported in the literature [1,7,8].

Test results indicate presence of multiple crack initiation and growth sites and periodic FCG with long intermittent dormant intervals as evidenced by AE and optical analysis. Both mode I (opening) and mode II (shearing) were observed, however, catastrophic failure consistently occurred under mode I. Crack growth activity was dominated by the propagation near the interface between the facesheets and the core that lead to weakening of the two phase action and subsequent cracking of the facesheets. Some similarities to the failure sequence reported in the literature were observed except that in the current work significant fiber rupture never took place until catastrophic failure [1,7,8]. In the literature, the endurance limit for sandwich composites has been reported to be as low as 60% of the ultimate static load [7], however, life was found to be unlimited below 75% in the current study. Energy dissipation as a result of multiple crack initiation and propagation sites may be responsible for the high endurance limit as it effectively reduces crack tip stress intensity and, therefore, stagnates crack tip(s) advancement. Core thickness effect is significant in sandwich composites. Preliminary results indicate reduction in lifetime as the thickness of foam core is increased for the same facesheet thickness.

Further studies in this area are currently underway.

3 STIFFNESS REDUCTION MODEL

Difficulties in FCG and lifetime assessment arise as a result of multiple crack fronts and large differences in elastic properties of the constituents (core, interface, facesheet) do not yield data suitable for the implementation of existing models. Therefore, an AE based stiffness reduction model was developed based on the overall AE activity during the fatigue test. The underlying assumption was that the extent of the damage (or cracking) in each constituent of the sandwich composite is directly proportional to the AE activity in that constituent. This is a reasonable assumption as AE activity can not take place unless a particular constituent of the sandwich composite suffers damage during fatigue. Additionally, Kaiser's effect prohibits replication of AE activity associated with a particular event [10].

This model was based on the percentage of total AE energy released by each constituent, which was taken as being proportional to the percentage of damage. This energy percentage, in turn, was determined from the overall number of events corresponding to each constituent of the sandwich composite within a given interval. An important parameter in this model is the weight factor, κ_i, for each constituent. κ was calculated from the static test results as the sum of the product of the total number of events for each constituent and respective energy level divided by the total energy during a given time interval. The weight factor contains information as to the role that each constituent plays in maintaining the integrity of the sandwich composite. Stiffness reduction parameter ".°(*t*)" under static or fatigue testing can then be represented as,

$$\Delta\varepsilon(t) = \sum_{1}^{n} \phi_i(t)\kappa_i \tag{1}$$

Where ϕ_i reflects cumulative percent damage in the core, interface and the facesheet for a given time (or #cycles) interval, respectively; and κ_i are the corresponding weight factors. This model is robust in its application as it can be implemented in real time while AE data is continuously collected for a component in service. Thus this model becomes an important tool for calculating remaining lifetime or loss of lifetime at a given instant in the fatigue life.

4 STOCHASTIC ANALYSIS

To evaluate large scatter observed in the fatigue lifetime data, stochastic analysis was performed utilizing three parameter Weibull distribution [11],

$$\ln\left[\ln\left(\frac{1}{SF(N)}\right)\right] = \alpha_s \ln(N) - \alpha_s \ln(\upsilon_s) \tag{2}$$

where the survival probability *SF(N)* is the probability that life will be longer than *N*, *N* is a random variable υ_s the scale parameter. no,s, the location parameter denoting minimum life was set equal to zero in the current analysis. Eq. 2 can be applied at any number of cycles to obtain the survival probability. Regression analysis plots of ln(ln(1/*SF*)) vs ln(*N*) were generated, a typical example is shown in Fig. 3. The accuracy of Fig. 3 increased as the number of lifetime data was increased.

Calculation of shape (α_s) parameter indicated substantial scatter in the lifetime data independent of the stress level. However, the scale parameter (υ_s) indicates a clear ascending trend as a function of stress level. The main advantage of using this survival probability method is its simplicity in ascertaining remaining lifetime information. To check the accuracy of the proposed Weibull distribution, Kolmogorov Smirnov (K-S) Goodness-of-Fit tests were performed. To perform this test, F*(Ni), the observed cumulative distribution histogram and F*(Ni), the hypothesized cumulative distribution function were evaluated at 5% confidence level. K-S test assured the proposed stochastic model to yield the results within 5% accuracy.

5 CONCLUSIONS

AE proved to be very robust in determining the extent and location of damage, that lead to lifetime stiffness reduction model. Core failure dominated the damage

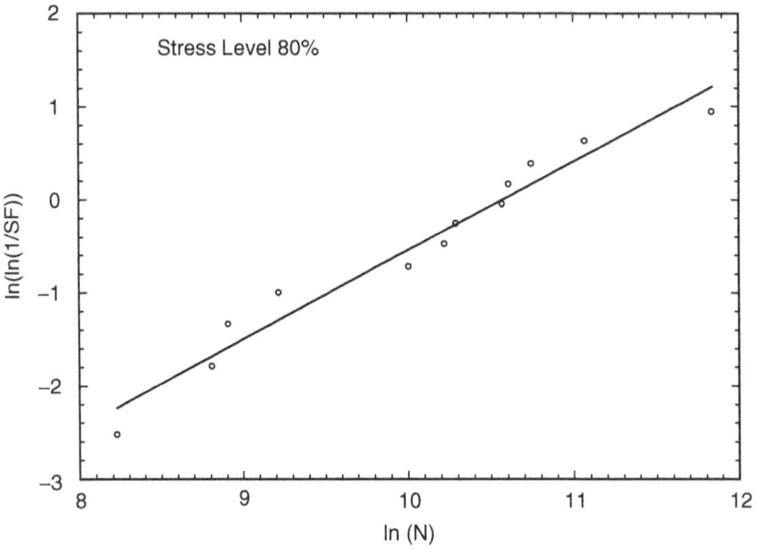

Figure 3: Typical result of regression analysis performed at 80% stress level.

mechanism, whereas, fiber rupture triggered the onset of catastrophic failure. Large scatter in fatigue lifetime and crack growth data was analyzed using Weibull distribution.

ACKNOWLEDGMENTS

The authors wish to acknowledge the authorities at ONR-Composites' for Marine Structures division for their financial support for this work. Special thanks are due to Dr. Yapa Rajapakse, the ONR program manager for his unrelenting support and guidance.

References

[1] Burman, M.: Fatigue Behavior Initiation and Propagation in Sandwich Structures, Royal Institute of Technology, Report No. 98-29, ISSN 0280-4646, 2000, Sweden.
[2] Bakukas, J., Prosser, W., Johnson, W.: Monitoring Damage Growth in Titanium Matrix Composites Using Acoustic Emission. Journal of Composite Material 1994:28(4):305–328.
[3] Harris, D., Dunegan, H.: Continuous Monitoring of Fatigue Crack Growth by Acoustic Emission Technique. Third SESA International Congress on Experimental Mechanics, 1974, Los Angeles.

[4] Li, F. and Li, Z. Acoustic Emission Monitoring of Fracture of Fiber-Reinforced Concrete in Tension. ACI Materials Journal 2000:97(6): 629–636. 5. Huang M. Using Acoustic Emission in Fatigue and Fracture Materials Research. JOM, 1998:50(11):1–14.

[6] English, L.: Listen and Learn, AE Testing Composites. Physical Acoustic Corporation, TR103-75-6/87, 1987, New Jersey.

[7] Mahi, A., Farooq, M., Sahraoui, D.: Mechanical behavior of Sandwich Composite Material under Cyclic Fatigue, New Trends in Fracture and Fatigue-Metz, 2002:April, France.

[8] Kulkarni, N., Mahfuz, H., Jeelani, S. and Carlsson, L. Fatigue Crack Growth and Life Prediction of Foam Core Sandwich Composite under Flexural Loading, Composite Structures 2003:59: 499–505.

[9] Caprino, G.: Predicting Fatigue Life of Composite Laminates Subjected to Tension-Tension Fatigue, Journal of Composite Materials 2000: 34(16):1334–1355.

[10] Nondestructive Testing Handbook, 2nd ed. American Society for Nondestructive Testing 5, ASNT, 1987, New Jersey.

[11] Sobzcyk, K., Spencer, B.: Random Fatigue: From Data to Theory, 1992, Academic Press, New York.

Earthquakes and Acoustic Emission – Carpinteri & Lacidogna (eds)
© 2007 Taylor & Francis Group, London, ISBN 978-0-415-44402-6

ON-PROCESS MONITORING OF CERAMICS AND CERAMIC COATINGS BY LASER AE

M. Enoki & S. Nishinoiri
Department of Materials Engineering, The University of Tokyo,
Tokyo 113-8656, Japan

ABSTRACT

Microcracks are easily generated during the processing in ceramic materials due to thermal stress because ceramics have intrinsic brittleness. It is very important to control microcracks or damages in ceramics during fabrication to ensure mechanical reliability of materials and structures. Especially it becomes necessary to avoid microcracks during manufacturing in large ceramic components which are fabricated for structural applications in recent years. Also many microcracks are easily introduced in thermal spraying of ceramic coatings which are widely used for heat resistant components. These cracks control mechanical and thermal properties of coating layer. Acoustic emission (AE) method is a powerful tool to monitor in-situ damage in materials because elastic waves are generated due to brittle cracking or frictions. Conventional AE techniques use a piezoelectric ceramic element and attach it to samples in order to detect AE signals at the surface of samples. Microfractures in materials are quantitatively analyzed by the detected waveforms such as location, size and fracture mode. However, conventional AE transducer cannot be used at elevated temperature or severe environment. Laser based ultrasonic (LBU) technique has been developed to characterize materials properties and detect flaws in materials. We have developed a non-contact AE technique to detect AE signals in various environments where conventional AE technique cannot be applied. AE during sintering of alumina ceramics and thermal spying of alumina powder on steel substrates were successfully measured by laser interferometers. The effect of processing parameters on AE behavior was clearly observed by analyzing AE waveforms. One of the most advantages of this laser AE technique is to estimate the temperature where microcracks are generated. These results could give a feed back to control processing conditions in order to avoid damage in materials. It was concluded that the laser AE method was very useful to detect microcracks in ceramics during fabrication.

1 INTRODUCTION

Recently, ceramics is extensively applied to structural materials because of its strength, abrasion resistance, chemical stability and so on, and ceramics components become larger and more complex. Ceramic components are produced by heating and densifing green compact at elevated temperature. These components are sometimes easily fractured during sintering due to large size and complex shape of ceramic components, so it becomes a problem that this fracture interrupts manufacturing process. It is important to control the crack initiation and propagation in green compact during sintering to resolve this problem. Furthermore, it is necessary to understand the evolution of mechanical properties during sintering to control the crack behavior [1, 2].

As the operation temperature of gas turbines has been rapidly increased to achieve high efficiency in recent years, materials are required improve their stability and mechanical properties at elevated temperature above 1773 K. Thermal barrier coating (TBC) with ceramic coating layer has been developed to shield heat from the outside and increase thermal stability of the surface [3]. As these coatings are subjected to both external stress and internal thermal stress constantly or periodically, the evaluation of mechanical properties and failure process of TBC at elevated temperature is desired to ensure the integrity. A health monitoring of structural components in service is important to assess the integrity of TBC, and a process control is also very important to provide uniform coating thickness and maximize the coating properties, as well as an estimation of the mechanical properties of coatings such as thermal shock resistance, thermal fatigue resistance and creep resistance.

Acoustic emission (AE) technique is a promising tool for reliability assessment of materials because it can monitor the generation and growth of microcrack in real time. However, conventional contact AE technique has a limit in application at elevated temperature, because a conventional piezoelectric transducer cannot be used above about 800 K. We have investigated the non-contact AE measurement technique using laser interferometer as a AE sensor [4–7]. This laser AE technique has several advantages such as non-contact measurement, absolute velocity measurement of AE signals, and applicability for severe environment. The purpose of this study is to investigate the influence of processing parameters on the generation and growth process of defects during fabrication process of ceramics by means of an in-situ monitoring system based on laser AE technique.

2 EXPERIMENTAL

2.1 In-situ monitoring during plasma spray coatings

Experimental set up of in-process monitoring system for plasma spraying is shown in Fig. 1. A sample was fixed to the stage equipped on a turntable in the

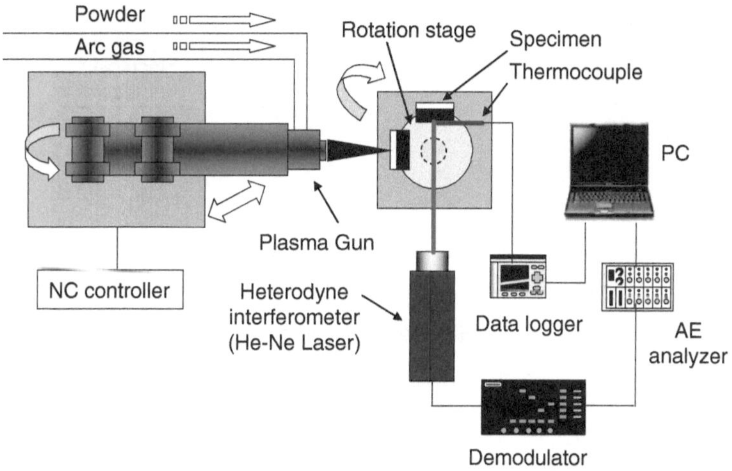

Figure 1: Schematic figure of in-process monitoring system using laser interferometer for plasma spraying.

direction perpendicular to the spraying direction. This coating process consists of three stages: First is a preheating of the substrate (stage-1), second is a spraying of topcoat (stage-2), and the last stage is an air cooling of the coated specimen (stage-3). After preheating of the surface by gun stroking without powder feeding, a feeding-rate control valve was opened to start the spraying of the topcoat. When the thickness of topcoat reached a desired value, the spraying apparatus was switched off and then the specimens were air-cooled until room temperature. After spraying, the sample stage was turned and the rear surface was faced to the laser beam of interferometer.

AE signals during cooling period were detected using a heterodyne type interferometer (AT-0022, Graphtec Corp.). Acquisition of AE signals was started after the power source of the spraying robot, exhaust duct and compressors were switched-off to avoid the influence of mechanical or electromagnetic noise on AE signals. Some dead time for AE measurement was required to adjust the focus of laser beam on the measuring surface. Total dead time before the start of an acquisition was about 1 min. Low noise type demodulator (AT-3600S, Graphtec Corp.) was used to measure an out-of-plane surface velocity on a sample with range of 1 mm/s/V. In order to reduce noise level, output signals were filtered with high pass filter (HPF) of 50 Hz and low pass filter (LPF) of 200 kHz. Detected AE waveforms were recorded by AE analyzer (DCM-140, JT-Toshi Corp.).

2.2 In-situ monitoring during sintering of alumina

Experimental setup for sintering of ceramics is shown in Fig. 2. Pre-treated alumina green compact, 35 by 35 by 20 m, was heated by electrical furnace, and

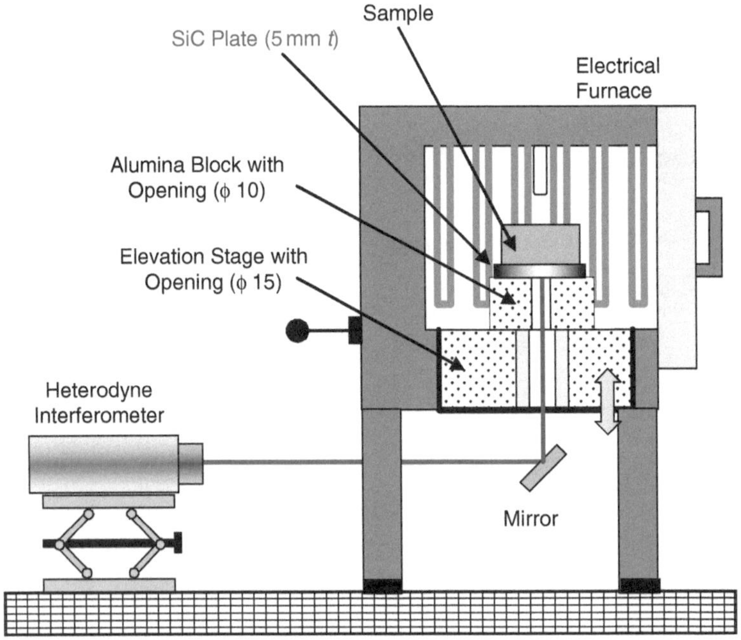

Figure 2: Schematic figure of in-process monitoring system using laser interferometer for sintering of ceramics.

AE signals during sintering were measured by using laser interferometers. Deep notch was introduced into some samples to investigate the effect of stress distribution. Two types of sintering pattern with high and low shrinkage rate were also used to change the stress during sintering. Refractory was holed to introduce a laser beam and AE signals through SiC plate was detected by the above measuring system.

Generally long waveguide is used to measure AE signals at elevated temperature in order to avoid the influence of heat on AE transducers. However, it is very difficult to characterize AE sources quantitatively because AE signals through waveguide are strongly affected by the shape of waveguide. AE signals detected by this set up have an advantage over the ones using waveguide.

3 RESULTS AND DISCUSSION

Figure 3 shows the temperature history and AE behavior for the different thickness of bond coat, where the other processing parameters were common, that is, pre-heating temperature Tp was 773 K, thickness of top coat d_{TC} was 1 mm, and gun velocity v was 0.1 m/s. In the sample without bond coat, only two AE with large amplitude were detected and immediately ceramic layer was delaminated.

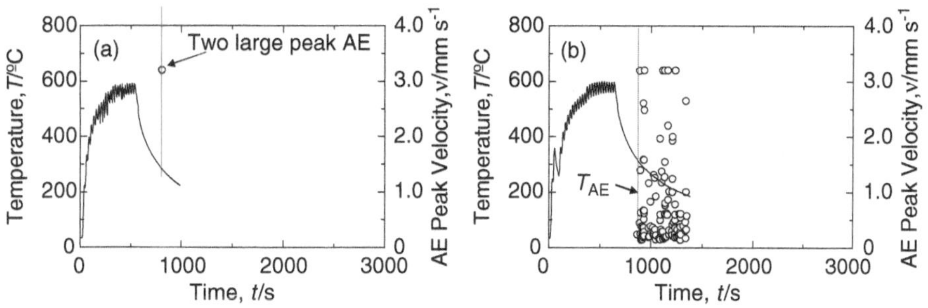

Figure 3: Relationship between temperature history and AE behavior for different thickness of bond coat, (a) $d_{BC} = 0\,mm$ and (b) $d_{BC} = 70\,\mu m$.

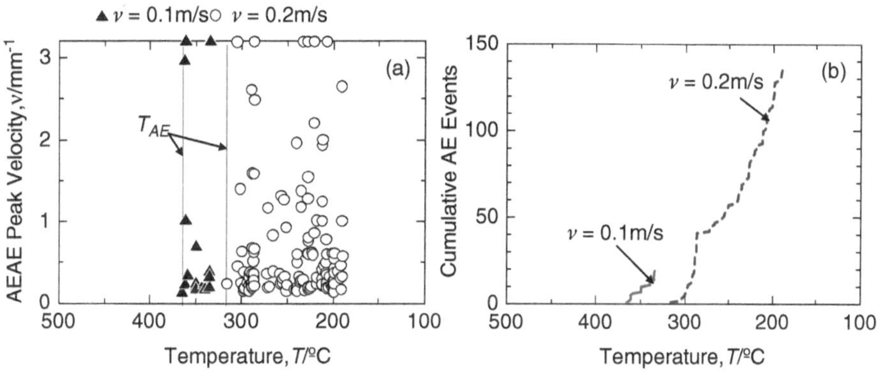

Figure 4: Relationship between the (a) amplitude or (b) cumulative AE event and generation temperature of AE for different traverse speed of spraying gun.

On the other hand, a large number of AE events with various amplitudes in the sample of thickness of bond coat d_{BC} of 70 μm were detected in the temperature range of 200 K before the delamination of top coat.

Figure 4 shows the relationship between the amplitude and generation temperature of AE for different traverse speed of gun, where d_{BC} was 70 μm, T_p was 773 K and d_{TC} was 1 mm. AE generation temperature T_{AE} in the sample of 0.1 m/s was higher compared with that in the sample of 0.2 m/s. The number of AE events in the case of 0.2 m/s was lager than that of 0.1 m/s, and the temperature range during AE generation of 0.2 m/s sample was wider than that of 0.1 m/s. Many interlamellar microcracks were observed in top coat of coating samples, and delamination was found near the interface between top coat and bond coat. It is difficult to quantitatively characterize the damage induced sample during coating. However, non-contact measurement of AE behavior enables to estimate the damage during coating process.

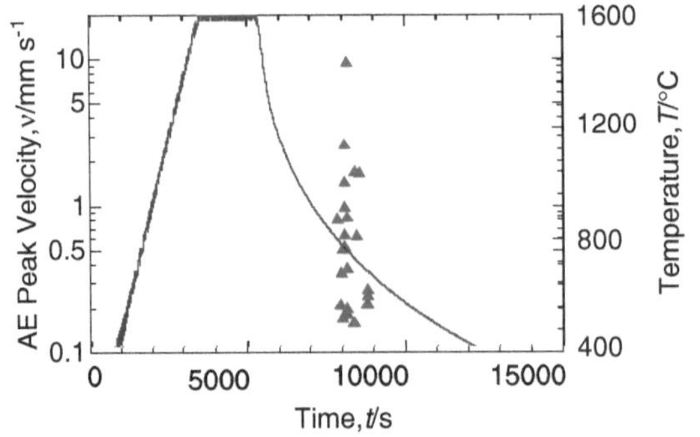

Figure 5: Typical AE behavior during sintering of alumina detected by laser AE technique.

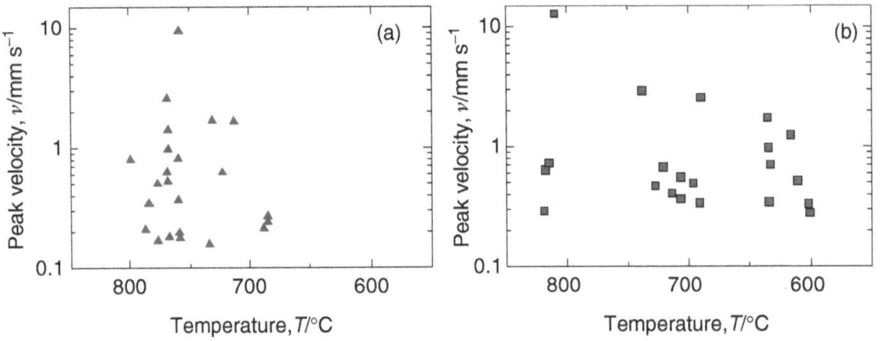

Figure 6: AE behavior during sintering, (a) smooth specimen and (b) notched specimen.

AE signals detected during cooling process could be classified into two types by the peak frequency, that is, Type-A and Type-B, respectively. Type-A signals have a peak around 75 kHz, on the other hand Type-B signals have several characteristic peaks in 100–200 kHz. Type-A signals were especially detected before the final delamination of specimens, and Type-B signals were broadly observed during cooling process.

Figure 5 shows a typical AE behavior in sintering of alumina, where AE generated at cooling period of processing process. Figure 6 shows the AE behaviors without and with notch. AE generation temperature of notch sample was higher than that of smooth sample, and AE continuously occurred until lower temperature. Figure 7 shows the effect of shrinkage rate of sintering on AE behaviors. AE generation temperature of high shrinkage rate, −0.171, sample was higher than that of low shrinkage, −0.143, rate one and many AE signals were detected in high shrinkage rate sample. These different AE behaviors clearly imply the stress distribution during sintering.

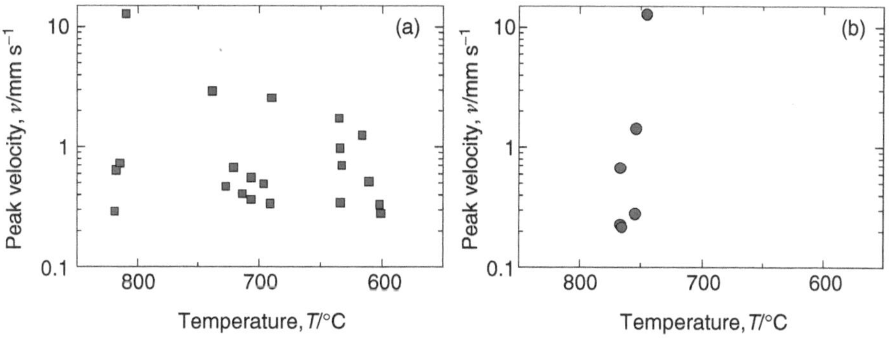

Figure 7: AE behavior during sintering, (a) high shrinkage rate and (b) low shrinkage rate.

4 SUMMARY

We developed a non-contact in-process monitoring system for ceramics and coatings with laser AE technique, and applied this system to the detection of microfracture during fabrication process. Conclusions of this study are as follows:

(1) Using the developed in-process monitoring system with laser AE technique, AE generated in cooling process of plasma spraying and alumina sintering was successfully detected.
(2) Processing conditions in plasma spraying affected AE behaviors such as generation temperature, number of AE and frequency characteristics.
(3) Difference in stress distribution due to notch and shrinkage rate during sintering of alumina also induced different AE generation temperature.

References

[1] S. Harato, T. Mitsudome, J. Hasegawa, Y. Hara, H. Matsumoto, T. Nose and M. Sugawara: Nippon Steel Technical Report, 59, 21–29 (1993).
[2] N. Miyata, T. Shiogai, C. Yamagishi and Y. Matsuo: J. Ceram. Soc. Jpn., 106, 494–499 (1998).
[3] R. A. Miller, J. Thermal Spray Technol. 6, 35–42 (1997).
[4] M. Enoki, M. Watanabe, P. Chivavibul and T. Kishi: Sci. Technol. Adv. Mater. 1, 157–165 (2000).
[5] M. Watanabe, T. Okabe, M. Enoki and T. Kishi: Sci. Technol. Adv. Mater. 4, 205–212 (2003).
[6] M. Watanabe, M. Enoki and T. Kishi: Mater. Sci. Eng. A359, 368–374 (2003).
[7] S. Nishinoiri, M. Enoki and K. Tomita: Mater. Trans. 45, 92–101(2004).

Earthquakes and Acoustic Emission – Carpinteri & Lacidogna (eds)
© 2007 Taylor & Francis Group, London, ISBN 978-0-415-44402-6

FRACTURING, ACOUSTIC EMISSION, AND NUMERICAL SIMULATION OF GRANITE UNDER MODE II LOADING

O. Stephansson[1], S. Stanchits[1], T. Backers[1,2], G. Dresen[1] & B. Shen[3]
[1]GeoForschungsZentrum GFZ Potsdam, Germany
[2]GeoFrames, Germany
[3]CSRIO, Australia

ABSTRACT

Mode II loading of drill core samples of medium-grained granite is conducted using the Punch-Through Shear (PTS-) test. Cylindrical samples with circular notches at both end surfaces are subjected to independent confining pressure and shear loading of the intact portion between the notches. This paper presents results from PTS- testing at confining pressure of 30 MPa, including the pressure dependency of Mode II fracture toughness, K_{IIC}, the resulting fracture pattern, acoustic emission (AE) characteristics and computer simulation. At failure a shear fracture connects the upper and lower notch. Fracture evolution on the macroscopic scale is described. Analysis of AE and micro-structural observations show contribution of both tensile and shear cracks to fracture propagation. AE polarity analysis shows a shift from dominantly tensile cracking to shear cracking during increasing punch loading. This suggests that fracturing in the PTS- test involves mixed-mode fracturing. The fracture initiation and propagation is analysed using the displacement discontinuity code FRACOD^{2D}. The code simulates the fracture initiation and propagation and acoustic emission of the PTS- test with a fair agreement with the experimental results.

1 INTRODUCTION

A new method for determination of the critical stress intensity factor in Mode II, the Punch-Through Shear (PTS-) test, was introduced by Backers [1]. The unique feature of the PTS- test is the ability to apply a confining pressure independent of the Mode II (shearing) load. This contribution reports the determined Mode II fracture toughness, K_{IIC}, and its dependency on confining pressure for Mizunami granite, and examines the AE (acoustic emission) activity and fracture content of the Mode II fracture. A two-dimensional fracture propagation code,

$FRACOD^{2D}$, has been applied to simulate fracture initiation and fracture propagation in PTS- samples. The results from modelling are compared to the resulting fracture pattern and AE recordings.

2 EXPERIMENTAL SETUP

The Punch-Through Shear (PTS-) test uses cylindrical samples with circular notches drilled centred into the end surfaces. The notches serve as friction free initial fractures. The experimental assembly is given in Fig. 1. A stiff servo-controlled (MTS) loading machine is used for axial loading. The inner cylinder of the sample is punched down at constant displacement rate ($3.3 \cdot 10^{-5}$ m/s) until failure, generating a localised shear stress between upper and lower notch. The applied confining pressure generates a normal stress on the shear fracture. Details on the Punch-Through Shear test are given in Backers et al. [1].

The acoustic monitoring system consists of eleven piezoelectric transducers glued to the sample surface. During testing, ultrasonic transmission tests are performed periodically to monitor P-wave velocities in different directions. Hypocenter location is determined by a least square iterative technique to an accuracy of ~2 mm. For details on the recording system and location analysis see Zang [2].

3 EXPERIMENTAL RESULTS

3.1 Fracture toughness and fracture content

The PTS- test is currently performed at confining pressures, P, up to 70 MPa. K_{IIC} increases non-linear with P (Fig. 1). Transition from steep to shallow slope is at

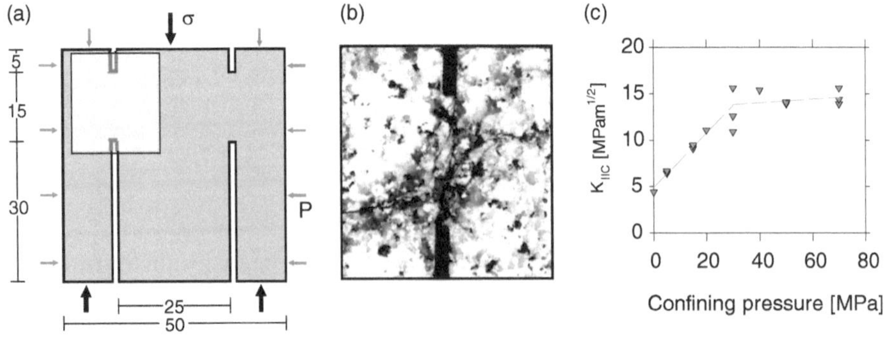

Figure 1: (a) Sample geometry, principle loading, and dimension for the PTS-Test in [mm]. (b) Cross cut view of the granite sample in the area between the notches as highlighted by the light grey area in (a). (c) K_{IIC} as function of the confining pressure, P, for Mizunami granite.

about 25–35 MPa. This was reported for several rock types (c.f. Backers et al. [3]). K_{IIC} of Mizunami granite rises from ~4.9 MPa $m^{1/2}$ at P = 0 MPa to ~14.9 MPa $m^{1/2}$ at P = 30 MPa, and shows ~15.1 MPa $m^{1/2}$ at P = 70 MPa.

The granite develops a wide process zone that is initiated at about 30% of the maximum load starting at the bottom notch at P = 30 MPa. In this intensely microcracked zone the main fracture develops and connects the notches at the peak load (Fig. 1). The rock develops a network of predominately grain boundary but also intragranular cracks. Typically features show the direction of en échelon fracturing (Backers et al. [4]).

3.2 Acoustic Emission

The fracture evolution for a sample loaded at P = 30 MPa as observed by AE analysis is shown in Fig. 2. (A) and (B). Continuous AE activity starts at about 45% of the failure load. The located events form clusters at top and bottom notch. With increase in axial load the bottom cluster starts to propagate upwards with approximately constant speed. The length of the AE process zone increases (~5–6 mm). The top cluster remains stationary (length: ~2–3 mm). After failure the events are evenly distributed between the notches at lower rate. For further details about the AE monitoring see Stanchits et al. [5].

AE first motion polarities were used to separate the signals into tensile (T- type), shear (S) and pore collapse (C) sources. Spatial distributions of T-S-C types of AE events at different stages of loading are presented in Fig. 2. (C). At the initial loading (t_1) equal amounts of T-, S- and C- type events are evident. Once the bottom cluster has started to move upwards (t_2), the relative amount of T- type events is lowered at bottom and top notch whilst the relative amount of C-type events is increased. The relative amount of S-type events remains constant. This picture is valid close to peak load (t_3) also. Post-peak the activity is lowered, but S- and C- type events are the dominant ones still.

4 MODELLING OF PTS TEST

FRACOD^{2D} (Shen [6]) is a two-dimensional code which is based on the Displacement Discontinuity Method (DDM) principles. It predicts the explicit fracturing process including fracture sliding/opening, fracture initiation and fracture propagation in rocks based on the F- criterion (Shen and Stephansson [7]). According to the F-criterion, in an arbitrary direction, è, at a fracture tip there exists a F-value, which is calculated by

$$F(\theta) = \frac{G_I(\theta)}{G_{Ic}} + \frac{G_{II}(\theta)}{G_{IIc}} \quad (1)$$

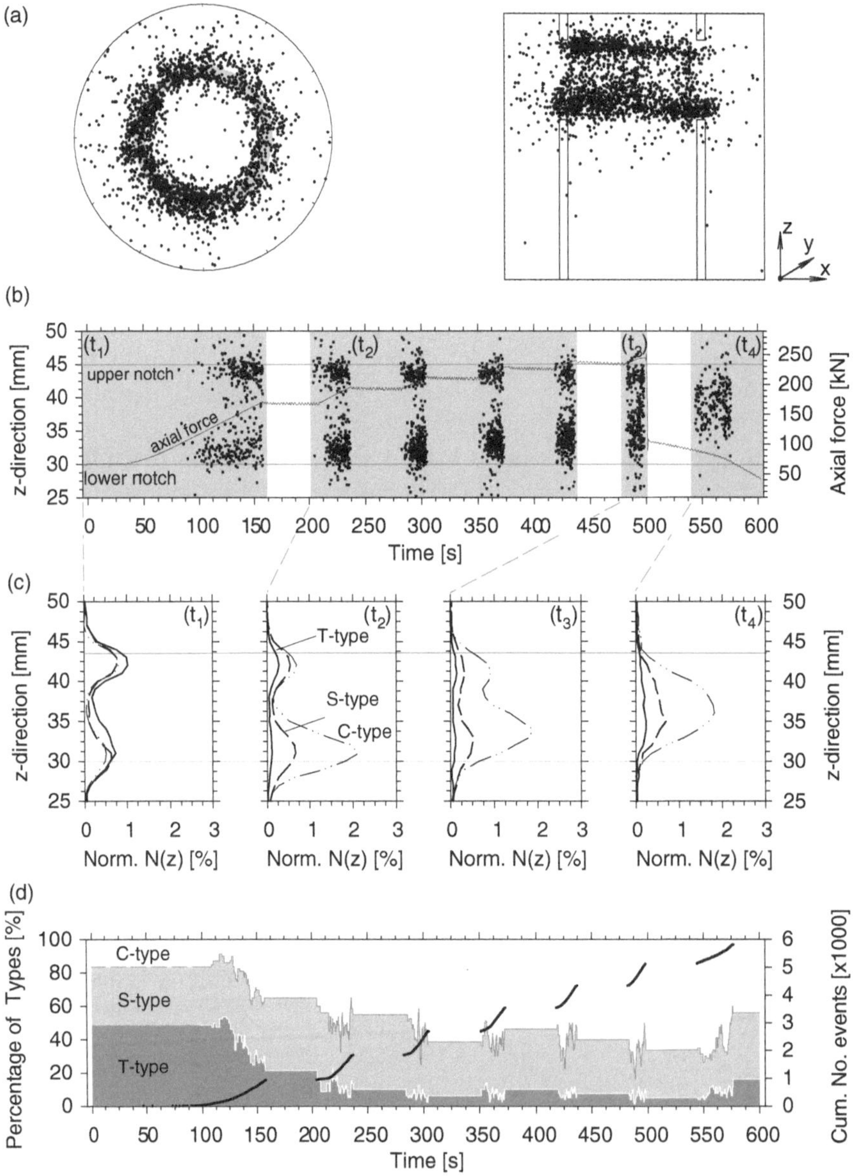

Figure 2: Acoustic Emission (AE) events in space and time. (a) AE distribution in map and side view. (b) The AE events are projected into z-direction and a force vs. time plot are given. (c) Polarity analysis in time slices as indicated in (a). (d) Time distribution of the Polarity distribution and cumulative AE number.

where G_{Ic} and G_{IIc} are the critical strain energy release rates for Mode I and Mode II fracture propagation; $G_I(\theta)$ and $G_{II}(\theta)$ are strain energy release rates due to the potential Mode I and Mode II fracture growth of a unit length. The direction of fracture propagation is defined as the direction of maximum F. At maximum F equal to 1.0, fracture propagation occurs.

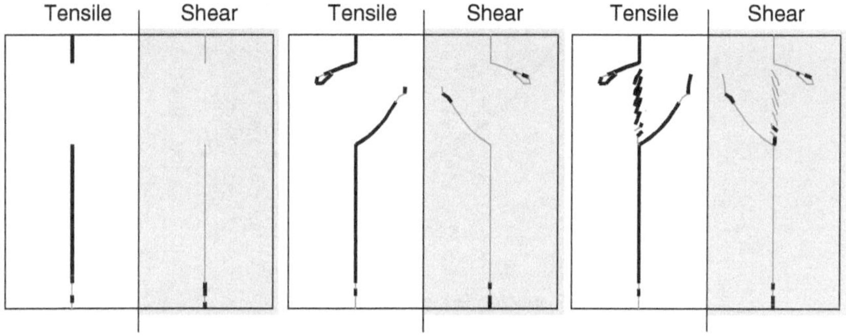

Figure 3: Fracture pattern as predicted from FRACOD modelling at different stages. Left side of each stage snap-shot shows tensile events while right hand side gives segments activated in shear.

$FRACOD^{2D}$ is applied to simulate the fracturing of the PTS- test of Mizunami granite. Input parameters to the code are: E = 49.9 GPa, $\nu = 0.37$, UCS = 166 MPa, tensile strength = 9 MPa, cohesion, c = 9 MPa, internal fraction angle, $\varphi = 33°$, $K_{IC} = 2.8$ MPa $m^{1/2}$, $K_{IIC} = 15.1$ MPa $m^{1/2}$, fracture normal stiffness $k_n = 13800$ GPa/m, fracture shear stiffness $k_s = 3099$ GPa/m, dilation angle = 0°, P = 30 MPa and axial load is 167 MPa.

The upper and lower notches are simulated by fractures with low strength and deformability. The fracture development during different stages is shown in Fig. 3. First a wing fracture is initiated at the bottom and top notches, but stops. Thereafter, a series of en échelon tensile fractures develops between the notches. Finally a set of shear fractures coalesces the tensile fractures and a main shear fracture is formed.

Fracture propagation as calculated by FRACOD is dominated by tensile fracture propagation in the pre-peak, i.e. pre-notch coalescence phase. Shear is evident only on reactivated segments of the wing fracture. During coalescence of the notches shear is the dominant mode, as the inclined struts formed in Mode I are connected by shear fractures.

5 COMPARISON AND DISCUSSION OF EXPERIMENTAL AND MODELLING RESULTS

The analysis of the fracture pattern and the history of fracturing on samples loaded to different fractions of the peak load gives a consistent picture of the fracture pattern. The shear loading in combination with the confining pressure, i.e. normal stress on the fracture, develops a wide zone of microcracking which connects to a shear fracture at peak load. Increasing the confining pressure increases the fracture toughness, but it does not show a linear increase as was demonstrated for K_{IC} (Winter [8]). The bi-linear increase might be dedicated to a change in dominant mechanism in the fracturing process. As K_{IC} and K_{IIC}

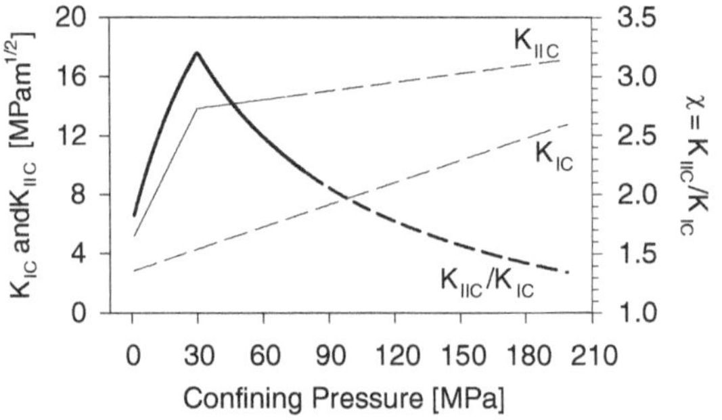

Figure 4: K_{IC}, K_{IIC} and the ratio $\chi = KI_{IC}/K_{IC}$ vs. confining pressure, P, for Mizunami granite. Assume $K_{IC} = K^0{}_{IC} + 0.05 \cdot P$ for the increase of Mode I fracture toughness with increase of confining pressure, as shown by Winter (1983) for Ruhr sandstone up to P = 100 MPa, extension of the linear trend for K_{IC} up to higher confining pressures, and low slope increase of K_{IIC} above the tested range of confining pressures. Dashed lines indicate assumed data. χ produces a maximum at a confining pressure of about 25–30 MPa.

show different behaviour to the increase of confining pressure (Fig. 4), presumably the Mode I fracturing on the microscale is increasingly suppressed leading to "purified" Mode II fracture propagation at elevated confining pressure.

The location analysis of the AE gives a very good description of the spatial distribution and history of the fracturing process. Despite the limited spatial resolution of about 2 mm the physical fracture pattern is very nicely mirrored by the AE. The polarity analysis indicates a distinct contribution of tensile dominated fracturing (T-type) at the notches on progressive loading. The relative amount of T-type events becomes smaller once the fracture starts propagating close to peak load. Hence it may be concluded that the tensile dominated events are predominately related to the formation of the early process zone. Increased shear load results in a shift to shear (S-type) and most prominently pore collapse (C-type) events. During the formation of the matured fracture process zone and the onset of fracture propagation more and more shearing and related pore collapse takes over, resulting in presumably shear enhanced compaction during the connection of the notches. Here C-type events contribute to more than 60%. The total contribution of the tensile dominated fracturing as defined by AE is very little compared to the S- and C- type events, which can both be linked to the shearing process.

While the fracture pattern and the Acoustic Emissions are both physically measurable, the FRACOD simulation is a simplified model. In contrast to the observed fracture pattern, the numerical modelling results show tensile fracture initiation and the formation of tensile wing fractures at the top and bottom notch, cf. Fig. 3. This feature is typically observed in experiments at low confining pressures ($P < 20-30$ MPa), but not at the given boundary conditions.

As the wing fracture comes to a halt in the numerical model, tensile en èchelon fractures form between the top and bottom notch. The recording of the AE also depicts this situation. As fracturing within the modelling proceeds the tensile fractures coalesce in shear and a major shear fracture is formed. At this stage the model becomes unstable.

According to the F-criterion as described in Section 4, tensile fracturing should dominate as G_{IIC} is in excess of G_{IC} or $K_{IIC} > K_{IC}$. To test the validity of the F-criterion of FRACOD the same model with the same material properties was run but now with $K_{IIC} < K_{IC}$. The result gives no tensile fracturing and wing crack formation but a straight through-going shear fracture between the upper and lower notch. This situation has also been simulated in PTS testing of porous mortar where K_{IIC} is less than K_{IC}. A sensitivity analysis about the influence of the model parameters has been performed and the results will be presented in the full paper of this contribution.

6 CONCLUSIONS

(1) Shear fracture initiation within the PTS- test are closely associated with formation of up to 50% of tensile cracks at the beginning of loading suggesting the formation of an initial process zone that is dominated by dilatant cracks.
(2) Pore collapse events increase at the expense of T-type events during progressive loading. This suggests that during shearing previously formed cracks are compacted.
(3) FRACOD code simulates the fracture initiation and propagation with a fair correspondence with the experimental results.

References

[1] Backers T., Stephansson O. & Rybacki E. Rock Fracture Toughness Testing in Mode II – Punch-Through Shear Test. Int. J. Rock Mech. Min. Sci, 39: 755-769. 2002.
[2] Zang, A., Wagner, F.C., Stanchits, S., Dresen, G., Andresen, R. & Haidekker, M. Source analysis of acoustic emissions in Aue granite cores under symmetric and asymmetric compressive loads. Geophysical Journal International; 135: 1113–1130. 1998.
[3] New data
[4] Backers T., Rybacki E., Alber M. & Stephansson O. Fractography of rock from the new Punch-Through Shear Test. In: Dyskin, A.V., Hu, X. & Sahouryeh, E. (eds.). Structural Integrity and Fracture – The International Conference on Structural Integrity and Fracture, Perth, Australia: 303–308. 2002.
[5] Stanchits S., Backers T., Stephansson O. & Dresen G. Comparison of Acoustic Emission Events and Micromechanics of granite under Mode I

and Mode II loading. In: Makurat A. & Curri P. (eds.). EURO-Conference on Rock Physics and Geomechanics – Micromechanics, Flow and Chemical Reactions. Extended Abstract Volume, 7–11 September 2003, Delft, Netherlands. 2003.

[6] Shen, B. FRACOD User's manual. Fracom Ltd. 2002.

[7] Shen, B. & Stephansson, O. Modification of the G-criterion of crack propagation in compression. Int. J. Eng. Fract. Mech.; 47: 177–189. 1993.

2.4 AE Structural monitoring and damage assessment

Earthquakes and Acoustic Emission – Carpinteri & Lacidogna (eds)
© 2007 Taylor & Francis Group, London, ISBN 978-0-415-44402-6

DAMAGE QUANTIFICATION FOR CONCRETE STRUCTURES BY IMPROVED *b*-VALUE ANALYSIS OF AE

T. Shiotani[1], X. Luo[2], H. Haya[2] & M. Ohtsu[3]
[1]Research Institute of Technology, Tobishima Corporation, 5472 Kimagase, Noda, Chiba·270-0222, Japan
[2]Railway Technical Research Institute, 2-8-38 Hikari-machi, Kokubunji, Tokyo 185-0034, Japan
[3]Graduate School of Science and Technology, Kumamoto University, Kumamoto 860-8555, Japan

ABSTRACT

Damage evaluation of concrete structures, in general, has been investigated visually. Except for seriously damaged structures, as a result, evaluation of moderately or intermediately damaged structures is usually difficult. Although prospective procedures using AE testing have already been proposed, the difference of monitoring condition in each application sometimes leads to non-quantitative results. Improved *b*-value is a slope of the peak amplitude distribution of AE signals. This value is known to be not influenced by the monitoring conditions, and to uniquely vary depending on the damage level. Accordingly, the improved *b*-value is applied to quantify the damage degree in damaged concrete piers due to an earthquake. Two sensor arrays with different damage levels are employed, and improved *b*-value based on the AE events located within the pier is obtained. As a result, the damage degree can be reasonably quantified by the improved *b*-value, which is promisingly employed as a damage index.

1 INTRODUCTION

By means of AE testing two indices have so far been used to inform the damage level. Felicity ratio (Fowler [1]) is one of those indices obtained from the stress level applied in comparison with the maximum stress level, and other is Calm ratio obtained from a ratio of accumulated number of AE activity during uploading to that during unloading. For the Felicity ratio, it is difficult to apply for in-situ monitoring since the maximum stress of which the materials have experienced is not readily estimated so that authors have proposed the RTRI instead (Luo et al. [2]). The RTRI is obtained on the basis of the maximum value during "a inspection period," and therefore the maximum value experienced has not been necessarily required in the calculation. Through the applications of those damage indices for in-situ concrete structures, however, several issues to be resolved still exist, for example, measurement of deformation is mandatory when obtaining the RTRI, moreover the deformation should exceed the minimal resolution of the value, otherwise it becomes difficult to obtain the onset of AE activity with respect to the deformation. Details discussion can be found in reference (Shiotani et al. [3,4]). Besides those indices, AE peak amplitudes are known to be closely related to the scale of fracture. Thus, the scale of peak amplitude might be larger with progress of fracture. However, it seems difficult only to pay attention to the peak amplitude. Because as fracture develops, an apparent mechanical property of the structure would change as well, in short, condition of media of which AE signals travel would be worse i.e., a high attenuation rate causes the fact that large scale of AE peak amplitude at the source results in small scale at the sensor. Thus, in the present paper the peak amplitudes are going to be studied as their distributions.

2 PEAK AMPLITUDE DISTRIBUTION

To analyze AE data acquired, a parameter analysis is widely and frequently employed. Among AE parameters, in particular, the AE peak-amplitude is considered a closely related parameter to the magnitude of fracture. Many researchers have thus studied *b*-value and *m*-value determined from a negative gradient of the peak-amplitude distribution. To apply the *b*-value analysis for fracture evaluation in slope failure, the method of calculating the *b*-value has been modified by incorporating statistical values of amplitude distribution (Shiotani et al. [5,6]). This is now referred to as improved *b*-value (I*b*-value) analysis.

2.1 Improved *b*-value

The range of AE amplitude in the improved *b*-value analysis is determined based on such statistical values as the mean μ and standard deviation α. Setting the accumulated amplitude over w_1 and w_2, as $N(w_1)$ and $N(w_2)$, I*b*-value is given by,

$$Ib = \frac{\log_{10} N(w_1) - \log_{10} N(w_2)}{(\alpha_1 + \alpha_2)\sigma}, \tag{1}$$

where α_1 and α_2 are constants.

The other modification in the I*b*-value analysis over the conventional *b*-value is the number of AE event analyzed. The number of AE data points is formulated by,

$$\int_0^{\infty} n(a)da = \beta, \tag{2}$$

where $n(a)$ is the number of AE hits (or events) at da and β is the total number of AE hits (or events) data subjected to the Ib-value analysis. Any number over 50 is considered appropriate as β. In consideration of the problem in population parameters, β from 50 to 100 would be an appropriate range to calculate the Ib-value. (Shiotani et al [5]). An appropriate number of β in actual concrete structures are further studied in the session followed. When a comparison between the seismic b-value and Ib-value is made, the Ib-value should be multiplied by 20.

3 IN-SITU APPLICATION

3.1 Structures and monitoring condition

AE monitoring was conducted in a RC pier of a railway rigid-frame bridge. In the pier, notable cracks were not found above the ground but seriously damaged condition was readily found after ground excavation. The crack condition is shown as in Fig. 1. In the north side, for example, a crack of 0.7 mm wide was found just above the footing and a diagonal crack, namely a shear type of crack, was observed at 1 m above the footing. Those cracks distributed up to 1.5 m above the footing. To know the AE activity not only for seriously damaged but intermediately damaged state, two arrays of AE sensors were employed i.e., A-array and B-array as shown in Fig. 2. A-array corresponds to seriously damaged

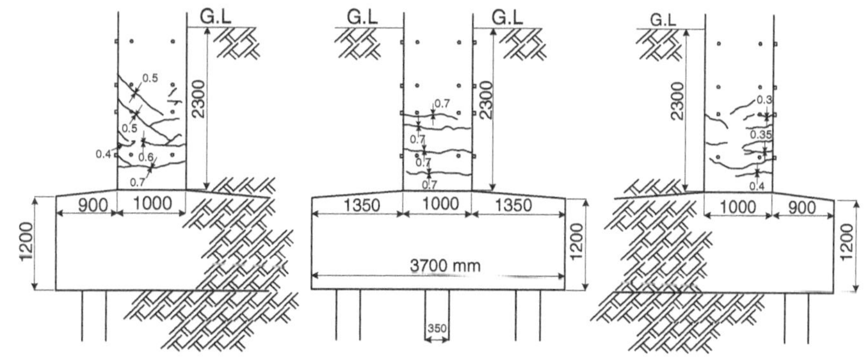

Figure 1: Crack condition of the RC pier, showing north, east, and south side from left to right.

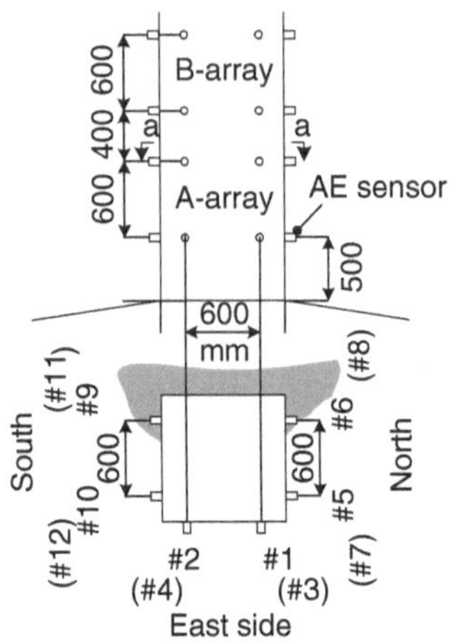

Figure 2: Two arrays of AE sensors.

state while B-array shows the intermediately damaged state. In each array, 12 AE sensors of 60 kHz resonance surrounds the pier. AE events induced by train passage were subsequently amplified by 40 dB at preamplifiers and fed to AE monitoring system (Mistras, Physical Acoustics Corp.) Both AE parameters and waveforms are recorded with the system.

3.2 Results

3.2.1 3D AE sources

AE sources induced by train passage are shown as in Fig. 3. 3D sources are all projected to the north sides plane, and the scale of the circle reflects the averaged peak amplitude of all hits contributing to the AE source/event. Fig. 3a was obtained from 10 times train passage, whereas Fig. 3b was from 4 times train passage, thus the number of sources, except for the scale of AE sources, could not be discussed here. A-array shows a lot of AE sources generated actively around the observed cracks (see Fig. 3a), and the scale of AE sources distributed widely, namely from small to large magnitude. In B-array, however, the large scale of AE sources as that in A-array could not be derived, and almost the identical scale of AE sources were presented.

3.2.2 Peak amplitude at AE sources

Based on the AE sources obtained in A-array and locations of AE sensors, attenuation rates were calculated as in Fig. 4. The negative coefficients only show the attenuation provided that AE energy decays as a function of propagation distance, but positive coefficients were presented in the chart i.e., not only attenuation but an amplification characteristic was obtained. This implies that it is difficult

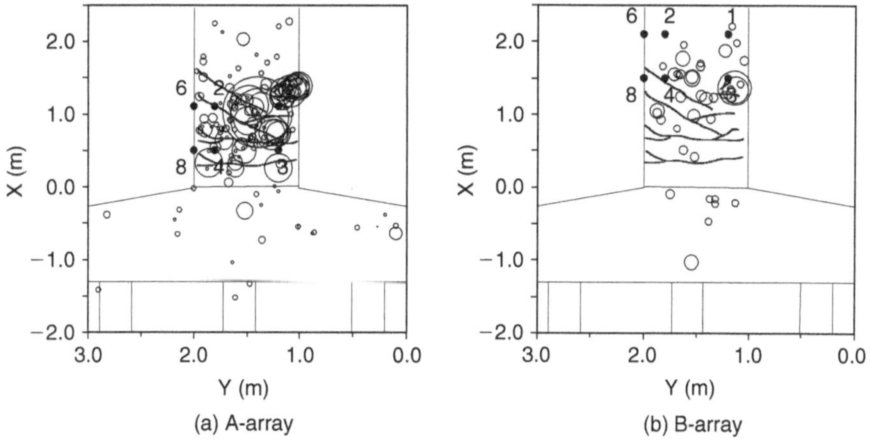

Figure 3: 3D AE source locations.

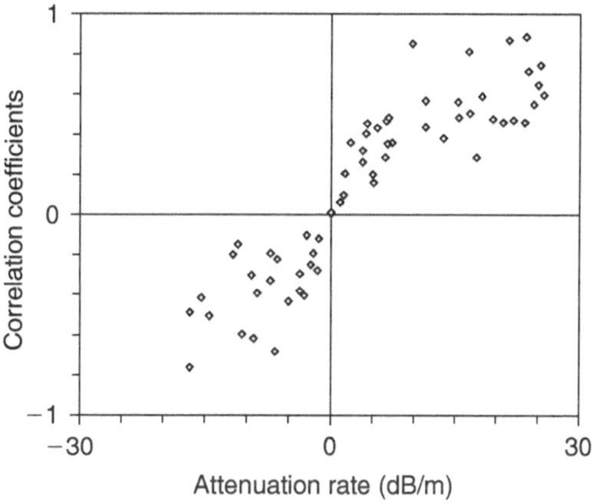

Figure 4: Attenuation rates.

to obtain attenuation rates in the materials of which cracks are randomly and intensively distributed. For example, in the case that a sensor was closely located to an AE source but a large number/scale of cracks was distributed in the vicinity of the sensor, resulting peak amplitude at the sensor would be smaller magnitude, while in the case that only a small number/scale of cracks was existed around the sensor, resulting peak amplitude at the sensor would be larger magnitude. This finding also suggests that in the structures of which heterogeneous crack-distribution from small to large scale are included, the peak amplitude at the source resulted in being difficult to calculate from obtained AE signals.

3.2.3 Peak amplitude distributions

AE peak amplitude distributions are exhibited as in Fig. 5. Open circles (B array) and open squares (A array) show the differential type of distribution and solid

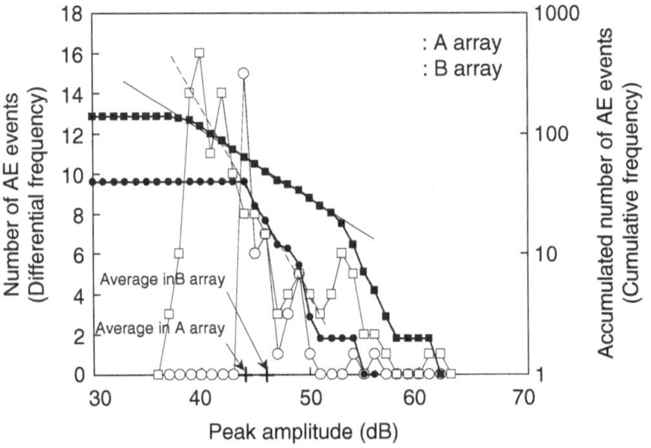

Figure 5: AE Peak amplitude distributions.

circles and solid squares show the cumulative type of distribution. In the figure, the averages of peak amplitudes in both arrays from all derived AE events are indicated with cross symbols as well. Again A-array corresponds to seriously damaged state while B-array shows the intermediately damaged state. The figure shows that the average of peak amplitude in B-array appeared larger value than that in A-array, suggesting that AE events generated from the intermediately damaged state had a larger value than seriously damaged state. Since, in general, the larger the scale of fracture becomes, the larger the AE peak amplitude has, the fact of the averages of peak amplitude does not suffice the general idea. In the differential type of distribution in B-array, AE events distributed narrowly around 45 dB, and AE events of larger than 50 dB did not appear so much, whereas it distributed widely from small scale to large scale of amplitude (see around 40 dB and 55 dB) in A-array. The difference of the differential distribution between both arrays resulted in thus the antithesis of general ideal. Solid plots show the cumulative types of distribution. In A-array, a descending curve with regard to the peak amplitude was linearly presented from 40 to 55 dB, where the gradient of the curve, when linear approximation is performed, are referred to as *b*-value. As in the same way, the cumulative distribution in B-array shows a decreasing trend as a function of peak amplitude, however, it can be readily found that the gradients would differ with the amplitude range employed. To resolve the difference in *b*-value depending on the range adopted, the improved *b*-value analysis ought to be performed (see Eqn 1). The improved *b*-value analysis was conducted for both distributions of A-array and B-array. A solid line shows the approximated line of A-array and a dotted line indicates that of B-array. The gradient in A-array (= 0.05) appeared smaller value than that in B-array (= 0.1), implying the improved *b*-value becomes smaller with damage levels.

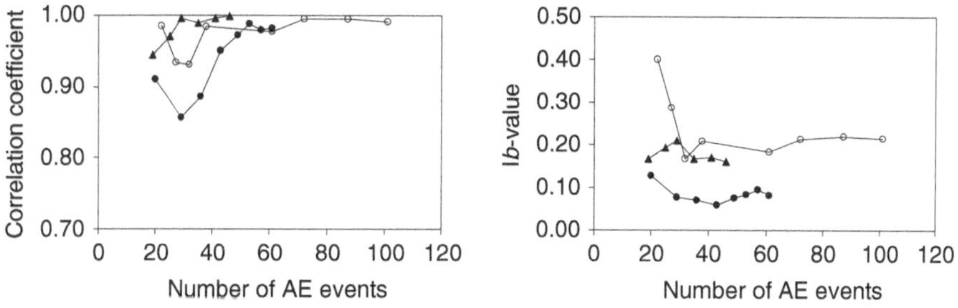

Figure 6: Correlation coefficients: left and improved *b*-value: right with the number of AE events.

4 DISCUSSION

4.1 Required number of AE data for the improved *b*-value analysis

To obtain the reasonable number of AE events in the improved *b*-value analysis, correlation coefficients and resulted improved *b*-values are shown with respect to the number of AE events as in Fig. 6. Three different populations of AE events, namely 46 (triangle), 61 (solid circle) and 101 (open circle), are shown. The correlation coefficients become more than 0.98 when AE events more than 40 are obtained (see Fig. 6 left), correspondingly improved *b*-value shows consistent when more than 40 events (see Fig. 6 right). This implies that the reasonable number of AE events contributing to the improved *b*-value analysis would be more than 40 events, in short, it is difficult to quantify the damage with the improved b-value from the number less than 40 events.

4.2 Improved *b*-value based on the results of one-dimensional source location

The improved *b*-values shown above were derived on the basis of 3D source locations, and therefore 3D source location seems to be necessarily performed when obtaining the improved *b*-value. Suppose in-situ damage investigation is performed with AE monitoring, installation of a large number of AE sensors into the structures might be in practical since it would take much time and cost. Fig. 7 shows the cumulative distribution of AE events when one dimensional source location was conducted. Due to small numbers of AE events in B-array, the chart only shows the result of A-array. The cumulative distributions were resulted from one dimensional sensor array with: ch-9 and ch-11 (68), ch-2 and ch-4 (42), and ch-5 and ch-7 (27) respectively, where the values in parentheses denote the AE events obtained. The resulting improved *b*-value from those three-combination of AE sensors, appeared smaller than 0.05 although the approximation becomes more difficult when the smaller number of events was

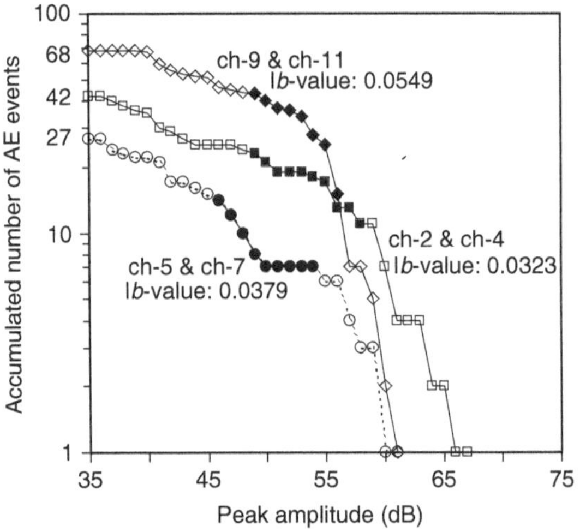

Figure 7: AE peak amplitude distributions in 1D source locations.

derived. Conclusively, the improved *b*-value analysis is possibly performed using a small number of AE sensors, for example, two sensors at minimum, however, the number of AE events contributing to the improved *b*-value analysis should always to be considered, i.e., as shown in Fig. 6, the AE events smaller than 40 might cause unstable/non- quantitative improved *b*-value, and the Fig. 7 accords well to the fact (see ch-5 & ch-7).

5 CONCLUSION

To quantify the damage of concrete structures, the improved *b*-value analysis, obtained from the AE amplitude distributions, was applied for the in-situ RC piers damaged. Two damaged areas with different damage levels were subjected to the AE monitoring. The results showed that it was difficult to discuss damage levels by only using peak amplitudes. Since the improved *b*-value appeared different values depending on the damage levels, the damage degree could be reasonably quantified with the improved *b*-value. Even with a small number of AE sensors, the improved *b*-value analysis was possible so that it will be potentially employed as a damage index in practical investigations.

References

[1] Fowler, T.J. "Experience with acoustic emission monitoring of chemical process industry vessels", Progress in Acoustic Emission III, JSNDI: 150–162, 1986.

[2] Luo, X., H. Haya, T. Inaba, T. Shiotani and Y. Nakanishi "Experimental study on evaluation of breakage in foundations using train-induced acoustic emission", Proc. Structural Engineering World Congress 2002: Paper No. T9-1-e-3, 2002.

[3] Shiotani, T., Y. Nakanishi, X. Luo, H. Haya and T. Inaba "Damage evaluation for railway structures by means of acoustic emission", Proc in APC-NDT 2003, Trans Tech Publications Ltd., Switzerland, in review.

[4] Shiotani, T., Y. Nakanishi, X. Luo, H. Haya and T. Inaba "Evaluation of structural integrity in railway structures using train-induced acoustic emission", Structural Faults and Repair 2003, Engineering Technics Press, (CD-ROM), 2003.

[5] Shiotani, T., K. Fujii, T. Aoki and K. Amou "Evaluation of progressive failure using AE sources and improved *b*-value on slope model test", Progress in Acoustic Emission VII, JSNDI: 529–534, 1994.

[6] Shiotani, T. and M. Ohtsu "Prediction of slope failure based on AE activity, Acoustic Emission: Standards and Technology Update", ASTM STP 1353:156–172, 1999.

Earthquakes and Acoustic Emission – Carpinteri & Lacidogna (eds)
 ISBN 978-0-415-44402-6

STRUCTURAL MONITORING AND LIFE-TIME ASSESSMENT OF MEDIEVAL TOWERS

A. Carpinteri & G. Lacidogna
Department of Structural and Geotechnical Engineering, Politecnico di Torino, 10129 Torino, Italy

ABSTRACT

The problem of structural stability of two medieval masonry towers is addressed. The geometrical and structural aspects of the towers were analysed and non-destructive tests were performed to assess the evolution of damage phenomena. The damage processes underway in some portions of the masonry were monitored using the Acoustic Emission (AE) technique. This method makes it possible to estimate the amount of energy released during the fracture process and to obtain information on the criticality of ongoing processes. Finally, an *ad hoc* theory based on fractal concepts for assessing the stability of masonry structures from the data obtained with the AE technique is proposed.

1 INTRODUCTION

Non-destructive and instrumental investigation methods are currently employed to measure and check the evolution of adverse structural phenomena, such as damage and cracking, and to predict their subsequent developments [1,2]. This study addresses some of the afore-mentioned problems deemed of special significance.

Towers geometry was defined through the customary survey methods. Damage, cracking and the evolution of these phenomena over time were assessed through a number of non-destructive techniques: thermographic exams were performed on the main sets of the towers subjected to cracking phenomena; tests with jack-flats were conducted in order to evaluate the range of stresses affecting the structures; at the same time, the cracking processes taking place in some portions of the masonry structures were monitored using the Acoustic Emission (AE) technique.

The AE technique has proved particularly effective [3,4], in that it makes it possible to estimate the amount of energy released during the fracture process and to obtain information on the criticality of the process underway. According to fractal concepts, an *ad hoc* theory is employed to monitor masonry structures by means of the AE technique [4]. The fractal theory takes into account the multiscale character of energy dissipation and the strong size effects associated with it. With this energetic approach it becomes possible to introduce a useful damage parameter for structural assessment based on a correlation between AE activity in a structure and the corresponding activity recorded on masonry elements of different sizes, tested to failure by means of double flat-jacks.

2 DESCRIPTION OF THE TWO TOWERS

These masonry buildings from the 13th century are the tallest and mightiest medieval towers preserved in Alba, a characteristic town in Piedmont (Italy).

Torre Sineo (Fig. 1) is square, 39 m high, and leans to a side by about 1%. Wall thickness ranges from 0.8 to 2 m. The bearing walls are *a sacco*, i.e., consist of brick faces enclosing a mixture of rubble and bricks bonded with lime and mortar. Over a height of 15 m, the tower is incorporated in a later building.

Torre Astesiano (Fig. 2) has a similar structure, but has a rectangular base. The filling material is more organised, with brick courses arranged in an almost

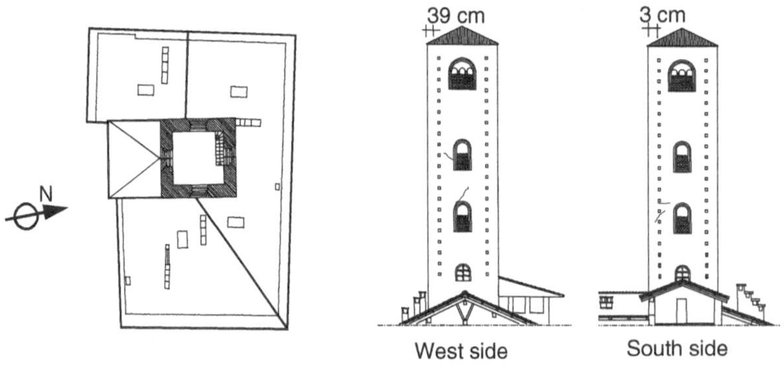

Figure 1: Torre Sineo. Plan and elevations of two sides of the tower. Notice the presence of cracks near the openings and the deviation from verticality of the tower.

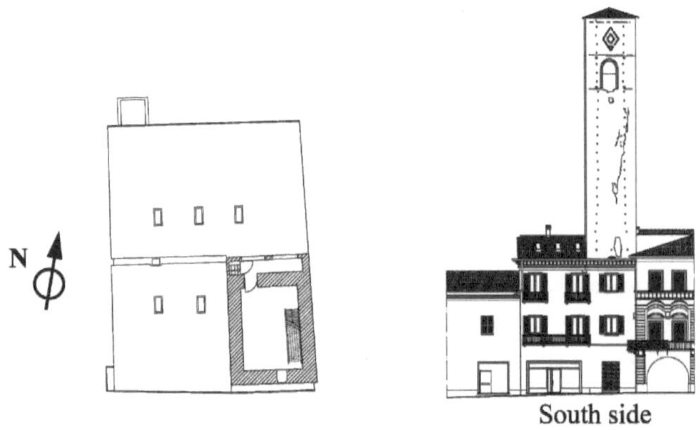

Figure 2: Torre Astesiano. Plan and elevation view of the tower. Notice the presence of the main crack in the upper part of the tower.

regular fashion, which, however, are not connected with the outer wall faces. In this case too, the total thickness of the masonry ranges from 2 m at the bottom to 0.8 m at the top. Total height is ca 36 m and the tower does not lean on any side. It is also incorporated in a later building, approx. 15 m high, built when the tower had been completed.

3 NON-DESTRUCTIVE EVALUATION TESTS

The geometry of the towers and the buildings they are embedded in was fully acquired and organized within a CAD system. The positions of the openings and the variations in the thickness of the tower walls was carefully recorded, together with the positions of the main cracks observed in the two structures. Tests with single and double flat-jacks were performed on the masonry walls of both towers. These tests were designed to estimate stress values in the masonry at different levels and to assess the elastic modulus and failure strength in situ.

Thermovision was used to identify areas with structural anomalies. Figure 3 shows a photograph of a badly damaged portion of the Sineo Tower (a) and compares it with the relative thermographic image (b). Some of the temperatures obtained are listed in the table: the coldest masonry points might reflect the presence of a crack. Figure 4 compares a thermographic image and a photograph of a damaged portion of the Torre Astesiano. From the thermographic image it can be seen that the average temperatures of the masonry zones in the proximity of the vertical crack are higher than the temperatures recorded at the crack. The reduction in temperature, in fact, reflects the gaps in the bricks along the crack.

The cracking processes taking place in some portions of the masonry structures were monitored using the Acoustic Emission (AE) technique. Crack opening, in fact, is accompanied by the emission of elastic waves which propagate within the bulk of the material. These waves can be captured and recorded by transducers applied to the surface of the structural elements [6].

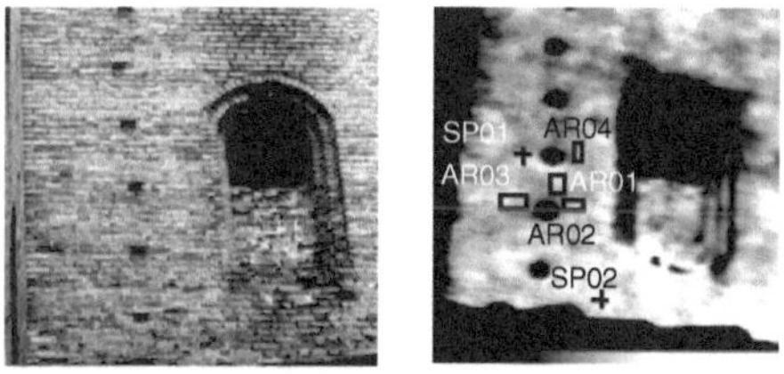

(a) (b)

Points	Temperature °C
Spot 1	30.2°C
Spot 2	32.7°C
AR 01 mean	31.6°C
AR 02 mean	30.8°C
AR 03 mean	31.6°C
AR 04 mean	30.8°C

Figure 3: Torre Sineo. Thermography: view of portion analysed (a). Temperature diagram in the 28.6–32.9°C range (b).

(a)

(b)

Points	Temperature °C
Spot 1	30.2°C
Spot 2	32.5°C
AR 01 mean	31.6°C
AR 02 mean	31.8°C

Figure 4: Torre Astesiano. Thermography: view of portion analysed (a). Temperature diagram in the 28.6–32.9°C range (b).

4 AE DAMAGE DETECTION OF THE TOWERS

For the Sineo Tower, through AE monitoring, two cracks were detected in the inner masonry layer at the seventh floor level (Fig. 1). The monitoring process revealed an on-going damaging process, characterized by slow crack propagation inside the brick walls. In the most damaged zone, crack spreading had come to a halt, the cracks having achieved a new condition of stability, leading towards compressed zones of the masonry. In this particular case it can be seen that, in the zone monitored, each appreciable crack advance is often correlated to a seismic event. In the diagram shown in Figure 5, the cumulative AE function relating to the area monitored is overlaid with the seismic events recorded in the Alba region during the same time period; the relative intensity of the events is also shown [4].

A similar behaviour was observed for the Torre Astesiano. This structure was monitored by means of two transducers applied to the inner masonry layer of the tower, at the fourth floor level near the tip of the large vertical crack. The results obtained during the monitoring period are summarised in the diagram in Figure 5. In this case too, it can be seen how the damage to the masonry and the propagation of the crack, as reflected by the cumulative number of EA events, evolved progressively over time. A seismic event of 4.7 degrees on the Richter scale occurred during the monitoring period: from the diagram we can see how the cumulative function of EA events grew rapidly immediately after the earthquake.

5 A FRACTAL CRITERION FOR AE MONITORING

Fragmentation theories have shown that during microcrack propagation energy dissipation occurs in a fractal domain comprised between a surface and the

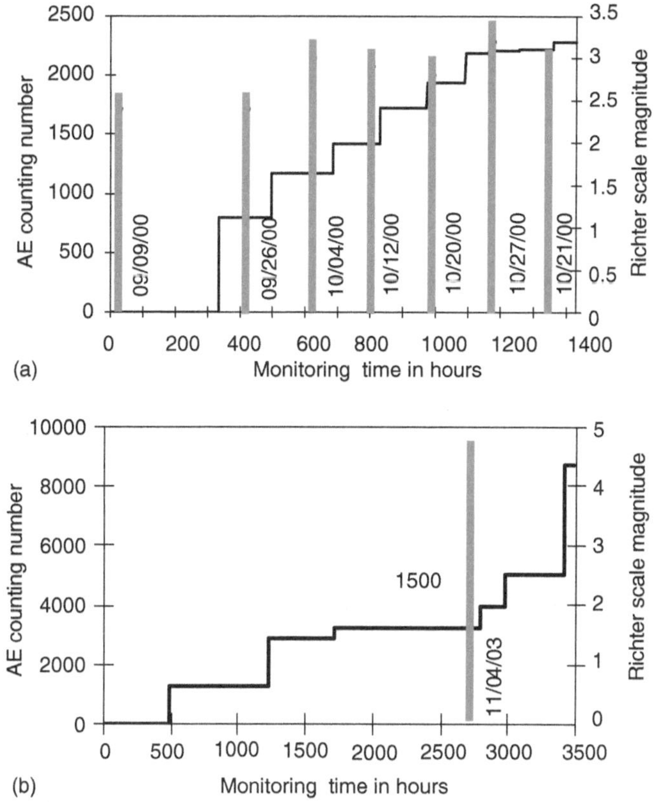

Figure 5: AE and seismic events. Torre Sineo (a), Torre Artesiano (b).

specimen volume V [5]. On the other hand, during microcrack propagation, acoustic emission events can be clearly detected. Since the energy dissipated, E, is proportional to the number of AE events, N, the critical density of acoustic emission events, Γ_{AE}, can be considered as a size-independent parameter:

$$\Gamma_{AE} = \frac{N_{max}}{V^{D/3}}, \tag{1}$$

where Γ_{AE} is the fractal acoustic emission density, N_{max} is evaluated at the peak-stress σ_u and D is the so-called fractal exponent comprised between 2 and 3. Eq. (1) predicts a volume-effect on the maximum number of AE events for a specimen tested to failure [4].

The extent of structural damage observed during the monitoring period, identified by the parameter η, can also be correlated to the rate of propagation of the microcracks. If we express the ratio between the cumulative number of AE events recorded during the monitoring process, N, and the number obtained at the end of the observation period, N_d, as a function of time, t, we get the damage time dependence on AE:

$$\eta = \frac{E}{E_d} = \frac{N}{N_d} = \left(\frac{t}{t_d}\right)^{\beta_t}. \tag{2}$$

In eq.(2), the values of E_d and N_d do not necessarily correspond to peak stress conditions ($E_d \leqslant E_{max}$; $N_d \leqslant N_{max}$) and the t_d parameter must be construed as the time during which the structure has been monitored. By working out the β_t exponent from the data obtained during the observation period, we can make a prediction as to the structure's stability conditions. If $\beta_t < 1$, the damaging process slows down and the structure evolves towards stability conditions, in as much as energy dissipation tends to decrease; if $\beta_t > 1$ the process becomes unstable, and if $\beta_t \cong 1$ the process is metastable, i.e., though it evolves linearly over time, it can reach indifferently either stability or instability conditions.

6 FLAT-JACK AND AE TESTS

In order to assess the extent of damage in the zone monitored using the AE technique, a compressive test was conducted on the masonry through the combined use of double jacks and AE sensors (Fig. 6). For the Astesiano Tower, compressive tests were performed on three different masonry sections at third level. The prismatic masonry volumes tested in compression were delimited crosswise by

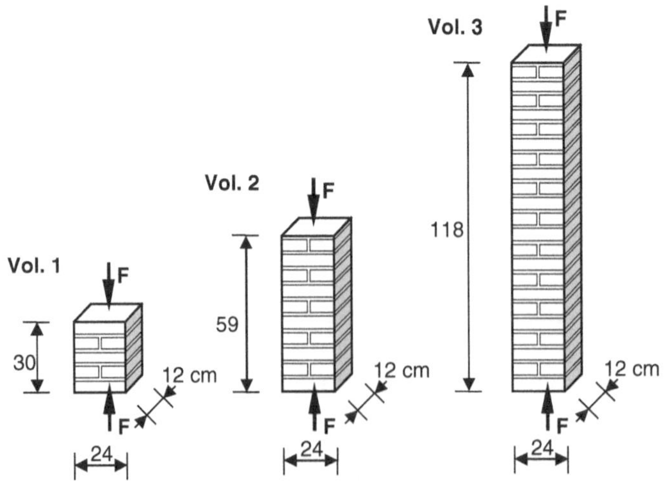

Figure 6: Masonry elements tested in compression by means of double flat-jacks and AE sensors.

Table 1: Experimental values obtained from flat-jack tests and AE measurements.

Specimen	Volume [cm^3]	Peak stress [MPa]	N_{max} at σ_u
Vol. 1	8640	2.07	~6500
Vol. 2	16992	1.61	~12000
Vol. 3	33984	1.59	~18000

vertical cuts (Fig. 6). The tests are in keeping with the procedures specified in ASTM 1991 [7], other than for the vertical cuts produced in order to eliminate, in the element damaged, the influence of the adjacent masonry portions. The experimental results are summarised in the Table 1. The table shows that in compressive tests the cumulative number of AE events increased with increasing specimen volume. From a statistical analysis of the experimental data, parameters D and Γ_{AE} (eq. (1)) can be quantified [4]. Parameter D represents the slope, in the bilogarithmic diagram, of the curve correlating N_{max} to specimen volume. By best-fitting, we obtain $D/3 \cong 0.743$, so that the fractal exponent, as predicted by fragmentation theories, turns out to be of between 2 and 3 ($D \cong 2.23$). Moreover, the critical value of fractal AE density turns out to be: $\Gamma_{AE} \cong 8.00\,cm^{-2.23}$.

7 DAMAGE LEVEL OF THE TOWERS

During the observation period, which lasted 60 days for the Sineo Tower and 146 days for the Astesiano Tower, the number of AE events recorded for the former was $N \cong 2250$, and for the latter it was $N \cong 9000$ (Fig. 5). Through earlier tests performed on rubble filled masonry, 80 cm thick, and hence characterised by appreciable discontinuities, it was ascertained that the transducers were able to pick up the AE signals from a distance of up to 10 m from their points of application and to a depth of 12 cm, i.e., over a length corresponding to the thickness of the outer layer of bricks.

Since the average width of the sides of the towers is ca 500 cm, the total volume monitored by the transducers will be: $V \cong 500 \times 2000 \times 12 = 1.2 \times 10^7\,cm^3$. From eq. (1), using fractal exponent $D \cong 2.23$ and the critical value of fractal acoustic emission density, $\Gamma_{AE} \cong 8.00\,cm^{-2.23}$, we obtain a critical AE number of $N_{max} \cong 1.46 \times 10^6$. Introducing the values of N_{max} into eq.(2), we get $\eta \cong 0.154\%$ for Torre Sineo and $\eta \cong 0.616\%$ for Torre Astesiano. These values represent, in percentage terms, the amount of energy released with respect to the energy that would cause the ultimate damage of the monitored volumes.

Finally, in order to obtain indications on the rate of growth of the damage process in the towers, as given in eq. (2), the data obtained with the AE technique were subjected to best-fitting in the bilogarithmic plane. For the Sineo Tower,

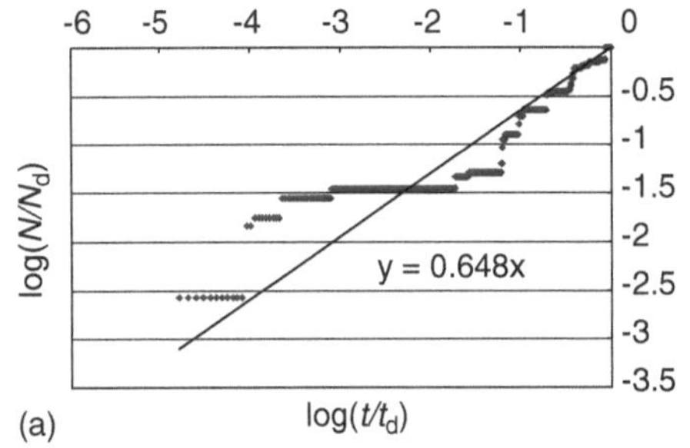

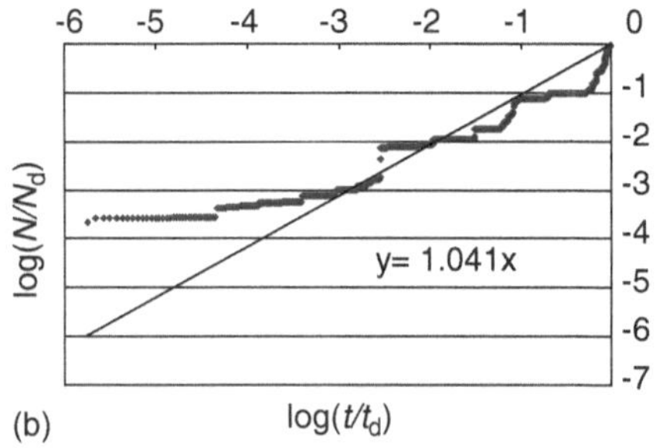

Figure 7: Evolution of damage: Torre Sineo (a), Torre Astesiano (b).

this yielded a slant $\beta_t \cong 0.648$, for the Artesiano Tower $\beta_t \cong 1.041$ (Fig. 7). These results confirm how the damage process stabilised in the Sineo Tower during the monitoring period, whereas for the Astesiano Tower it evolved towards a condition of instability according to a quasi-linear progression over time. In fact, if we introduce the values of N and N_{max} obtained for Torre Artesiano into eq. (2), with $\beta_t = 1.041$, we get $t/t_{max} \cong 7.532 \times 10^{-3}$. The life-time of this structure is therefore defined, in terms of time before the maximum number of AE events is reached in the analysed zone, at about 53 years.

8 CONCLUSIONS

In view of the appreciable number of old structures still in use today, more attention should be paid to preservation and rehabilitation issues. A sound safety assessment should take into account the evolution and the interaction of different damage phenomena. In this connection, the AE monitoring can be highly effective. This technique makes it possible to introduce a useful energy-based damage parameter for structural assessment which establishes a correlation between AE activity in a structure and the corresponding activity recorded on specimens taken from the structure and tested to failure. Moreover, by applying compressive tests through the combined use of double flat jacks and AE sensors, the safety of structures undergoing damage and degradation processes can be efficiently evaluated in situ.

ACKNOWLEDGEMENTS

The present research was carried out with the financial support of the Ministry of University and Scientific Research (MIUR) and of the European Union (EU). The authors would like to thank Architects M. Aprile and L. Bacco for the technical support provided in the structural monitoring.

References

[1] Carpinteri, A., Bocca, P.: *Damage and Diagnosis of Materials and Structures*, Pitagora Editrice: Bologna (1991).

[2] Anzani, A., Binda, L., Mirabella Roberti, G.: The effect of heavy persistent actions into the behaviour of ancient masonry, *Materials and Structures*, Vol. 33, pp. 251–261 (2000).

[3] Carpinteri, A., Lacidogna, G.: Damage diagnosis in concrete and masonry structures by acoustic emission technique. *Journal Facta Universitatis*, Vol. 3, pp. 755–764 (2003).

[4] Carpinteri, A., Lacidogna, G., Pugno, N.: Damage diagnosis and life-time assessment of concrete and masonry structures by an acoustic emission technique. Eds. V. C. Li, C. K. Y. Leung, K. J. Willam, S. L. Billington, pp. 31–40; *Proc. of 5th Intern. Conf. on Fracture Mechanics of Concrete and Concrete Structures (FraMCos-5)*. Vail, Colorado-USA (2004).

[5] Carpinteri, A., Pugno, N.: Fractal fragmentation theory for shape effects of quasi-brittle materials in compression. *Magazine of Concrete Research*, Vol. 54, pp. 473–480 (2002).

[6] Holroyd, T.: *The Acoustic Emission & Ultrasonic Monitoring Handbook*, Coxmoor Publishing Company's, Oxford (2000).

[7] ASTM: Standard test method for in situ compressive stress within solid unit masonry estimated using flat-jack measurements. ASTM C1196-91. Philadelphia (1991).

Author Index